The Mathematics Coach Handbook

Problem Solving in Mathematics and Beyond

Print ISSN: 2591-7234
Online ISSN: 2591-7242

Series Editor: Dr. Alfred S. Posamentier
Professor Emeritus of Mathematics Education, and Former Dean
CCNY - City University of New York

There are countless applications that would be considered problem solving in mathematics and beyond. One could even argue that most of mathematics in one way or another involves solving problems. However, this series is intended to be of interest to the general audience with the sole purpose of demonstrating the power and beauty of mathematics through clever problem-solving experiences.

Each of the books will be aimed at the general audience, which implies that the writing level will be such that it will not engulfed in technical language — rather the language will be simple everyday language so that the focus can remain on the content and not be distracted by unnecessarily sophiscated language. Again, the primary purpose of this series is to approach the topic of mathematics problem-solving in a most appealing and attractive way in order to win more of the general public to appreciate his most important subject rather than to fear it. At the same time we expect that professionals in the scientific community will also find these books attractive, as they will provide many entertaining surprises for the unsuspecting reader.

Published

Vol. 9 *The Mathematics Coach Handbook*
by Alfred S. Posamentier and Stephen Krulik

Vol. 8 *Tools to Help Your Children Learn Math: Strategies, Curiosities, and Stories to Make Math Fun for Parents and Children*
by Alfred S. Posamentier, Gavrielle Levine, Aaron Lieberman and Danielle Sauro

Vol. 7 *A Central European Olympiad: The Mathematical Duel*
by Robert Geretschläger, Józef Kalinowski and Jaroslav Švrček

Vol. 6 *Teaching Children to Love Problem Solving*
by Terri Germain-Williams

Vol. 5 *Strategy Games to Enhance Problem-Solving Ability in Mathematics*
by Alfred S. Posamentier and Stephen Krulik

For the complete list of volumes in this series, please visit www.worldscientific.com/series/psmb

Problem Solving in Mathematics and Beyond | Volume 09

The Mathematics Coach Handbook

Alfred S. Posamentier
City University of New York, USA

Stephen Krulik
Temple University, USA

World Scientific

NEW JERSEY • LONDON • SINGAPORE • BEIJING • SHANGHAI • HONG KONG • TAIPEI • CHENNAI • TOKYO

Published by

World Scientific Publishing Co. Pte. Ltd.
5 Toh Tuck Link, Singapore 596224
USA office: 27 Warren Street, Suite 401-402, Hackensack, NJ 07601
UK office: 57 Shelton Street, Covent Garden, London WC2H 9HE

Library of Congress Cataloging-in-Publication Data
Names: Posamentier, Alfred S., author. | Krulik, Stephen, author.
Title: The mathematics coach handbook / Alfred S. Posamentier and Stephen Krulik.
Description: New Jersey : World Scientific, [2018] | Series: Problem solving in mathematics and beyond ; Volume 9
Identifiers: LCCN 2018026804 | ISBN 9789813271708 (hardcover : alk. paper)
Subjects: LCSH: Mathematics teachers--Training of--United States. | Mathematics--Study and teaching--United States. | Mathematics--Vocational guidance--United States.
Classification: LCC QA10.5 .P68 2018 | DDC 510.71/2--dc23
LC record available at https://lccn.loc.gov/2018026804

British Library Cataloguing-in-Publication Data
A catalogue record for this book is available from the British Library.

For any available supplementary material, please visit
https://www.worldscientific.com/worldscibooks/10.1142/11028#t=suppl

We dedicate this book to our future generations so that they will be among the multitude that we hope will learn to love mathematics for its power and beauty!

To our children and grandchildren, whose future is unbounded

Lisa, Daniel, David, Lauren, Max, Samuel, Jack and Charles.

— Alfred S. Posamentier

Nancy, Dan, Jeff, Amy, Amanda, Ian, Sarah, and Emily.

— Stephen Krulik

Contents

About the Authors — *ix*

Introduction — *xvii*

Chapter 1 Math Coaches and Math Teachers — 1

Chapter 2 Coaching Teachers to Effectively Motivate Their Classes — 7

Chapter 3 The Art of Questioning — 33

Chapter 4 The Role of Problem Solving — 61

Chapter 5 Enriching Mathematics Instruction — 97

Chapter 6 The Various Roles of a Math Coach — 129

Chapter 7 Some Final Words — 151

About the Authors

Alfred S. Posamentier is Professor Emeritus of Mathematics Education at The City College of the City University of New York, and former Dean of the School of Education, where he was for 40 years. Previously, he was the Executive Director for Internationalization and Sponsored Programs at Long Island University, New York. Before that, he was Dean of the School of Education and Professor of Mathematics Education at Mercy College, New York. He is the author and co-author of more than 60 mathematics books for teachers, secondary and elementary school students, and the general readership. Dr. Posamentier is also a frequent commentator in newspapers and journals on topics relating to education.

After completing his B.A. degree in mathematics at Hunter College of the City University of New York, he took a position as a teacher of mathematics at Theodore Roosevelt High School (Bronx, New York), where he focused his attention on improving the students' problem-solving skills and at the same time enriching their instruction far beyond what the traditional textbooks offered. During his six-year tenure there, he also developed the school's first mathematics teams (both at the junior and senior level). He is still involved in working with mathematics teachers and supervisors, nationally and internationally, to help them maximize their effectiveness.

Immediately upon joining the faculty of the City College in 1970 (after having received his master's degree there in 1966), he began to develop inservice courses for secondary school mathematics teachers, including such special areas as recreational mathematics and problem solving in mathematics. As Dean of the City College School of Education for 10 years, his scope of interest in educational issues covered the full gamut educational issues. During his tenure as dean he took the School from the bottom of the New York State rankings to the top with a perfect NCATE accreditation assessment in 2009.

In 1973, Dr. Posamentier received his Ph.D. from Fordham University (New York) in mathematics education and has since extended his reputation in mathematics education to Europe. He has been visiting professor at several European universities in Austria, England, Germany, Czech Republic, and Poland, while at the University of Vienna he was Fulbright Professor (1990).

In 1989 he was awarded an *Honorary Fellow* at the South Bank University (London, England). In recognition of his outstanding teaching, the City College Alumni Association named him *Educator of the Year* in 1994, and in 2009. New York City had the *day*, May 1, 1994, named in his honor by the President of the New York City Council. In 1994, he was also awarded the *Grand Medal of Honor* from the Republic of Austria, and in 1999, upon approval of Parliament, the President of the Republic of Austria awarded him the title of *University Professor of Austria.* In 2003 he was awarded the title of *Ehrenbürger* (Honorary Fellow) of the Vienna University of Technology, and in 2004 was awarded the *Austrian Cross of Honor for Arts and Science, First Class* from the President of the Republic of Austria. In 2005 he was inducted into the *Hunter College Alumni Hall of Fame*, and in 2006 he was awarded the prestigious *Townsend Harris Medal* by the City College Alumni Association. He was inducted into the New York State *Mathematics Educator's Hall of*

Fame in 2009, in 2010 he was awarded the coveted *Christian-Peter-Beuth Prize* in Berlin, and in 2017 he received the *Summa Cum Laude nemine discrepante* Award from Fundacion Sebastian, A.C. in Mexico City.

He has taken on numerous important leadership positions in mathematics education locally. He was a member of the New York State Education Commissioner's Blue Ribbon Panel on the Math-A Regents Exams, and the Commissioner's Mathematics Standards Committee, which redefined the Standards for New York State, and he also served on the New York City schools' Chancellor's Math Advisory Panel.

Dr. Posamentier is a leading commentator on educational issues and continues his long time passion of seeking ways to make mathematics interesting to both teachers, students and the general public — as can be seen from some of his more recent books: *The Mathematics of Everyday Life* (Prometheus, 2018), *The Joy of Mathematics: Marvels, Novelties, And Neglected Gems That Are Rarely Taught in Math Class* (Prometheus, 2017), *Strategy Games to Enhance Problem-Solving Ability in Mathematics* (World Scientific, 2017), *The Circle: A Mathematical Exploration Beyond the Line* (Prometheus, 2016), *Problem-Solving Strategies in Mathematics: From Common Approaches to Exemplary Strategies* (World Scientific, 2015), *Effective Techniques to Motivate Mathematics Instruction* (Routledge, 2016) *Numbers: Their Tales, Types and Treasures* (Prometheus, 2015), *Teaching Secondary Mathematics: Techniques and Enrichment Units,* 9th Ed. (Pearson, 2015), *Mathematical Curiosities: A Treasure Trove of Unexpected Entertainments* (Prometheus, 2014), *Geometry: Its Elements and Structure,* (Dover, 2014), *Magnificent Mistakes in Mathematics* (Prometheus Books, 2013), *100 Commonly Asked Questions in Math Class: Answers that Promote Mathematical Understanding, Grades 6–12* (Corwin, 2013), *What Successful Math Teachers Do: Grades 6–12* (Corwin 2006, 2013), *The Secrets of Triangles: A Mathematical Journey* (Prometheus Books, 2012), *The Glorious Golden Ratio* (Prometheus Books, 2012), *The Art of Motivating Students for Mathematics*

Instruction (McGraw-Hill, 2011), *The Pythagorean Theorem : Its Power and Glory* (Prometheus, 2010), *Mathematical Amazements and Surprises: Fascinating Figures and Noteworthy Numbers* (Prometheus, 2009), *Problem Solving in Mathematics: Grades 3-6: Powerful Strategies to Deepen Understanding* (Corwin, 2009), *Problem-Solving Strategies for Efficient and Elegant Solutions, Grades 6–12* (Corwin, 2008), *The Fabulous Fibonacci Numbers* (Prometheus Books, 2007), *Progress in Mathematics* K-9 textbook series (Sadlier-Oxford, 2006–2009), *What Successful Math Teachers Do: Grades K-5* (Corwin 2007), *Exemplary Practices for Secondary Math Teachers* (ASCD, 2007), *101+ Great Ideas to Introduce Key Concepts in Mathematics* (Corwin, 2006), *π, A Biography of the World's Most Mysterious Number* (Prometheus Books, 2004), *Math Wonders: To Inspire Teachers and Students* (ASCD, 2003), and *Math Charmers: Tantalizing Tidbits for the Mind* (Prometheus Books, 2003).

Stephen Krulik is Professor Emeritus of Mathematics Education at Temple University in Philadelphia. While at Temple University Dr. Krulik was responsible for the undergraduate and graduate preparation of mathematics teachers for grades K-12, as well as the inservice alertness training of mathematics teachers at the graduate level. He teaches a wide variety of courses, among them, the History of Mathematics, Methods of Teaching Mathematics, and the Teaching of Problem Solving. This latter course grew out of his interest in problem solving and reasoning in the mathematics classroom. His concern that students understand the beauty and value of problem solving as well as the ability to reason led to his interest in problem solving.

Dr. Krulik received his B.A. degree in mathematics from Brooklyn College of the City University of New York, and his M.A. and Ed.D. in mathematics education from Columbia University's Teachers College. Before coming to Temple University, he taught mathematics in the New York City public schools for 15 years. At Lafayette High School in Brooklyn, he created and implemented several courses designed to prepare students for the SAT examination, while stressing the art of problem solving as opposed to rote memory of algorithms.

Nationally, Dr. Krulik has served as a member of the committee responsible for preparing the *Professional Standards for Teaching Mathematics* of the National Council of Teachers of Mathematics. He was also the editor of the NCTM's 1980 Yearbook *Problem Solving in School Mathematics.* Regionally, he served as president of the Association of Mathematics Teachers of New Jersey, was a member of the editorial team that produced the 1993 publication, *The New Jersey Calculator Handbook*, and was the Editor for their 1997 monograph, *Tomorrow's Lessons.*

His major areas of interest are the teaching of problem solving and reasoning, materials for teaching mathematics, as well as comprehensive assessment in mathematics. He is the author and co-author of more than 30 books for teachers of mathematics, including the *Roads to Reasoning* (Grades 1–8) and *Problem Driven Math* (Grades 3–8). Dr. Krulik is also the senior problem-solving author for a basal textbook series. Dr. Krulik is a frequent contributor to the professional journals, in mathematics education. He has served as a consultant to, and has conducted many workshops for, school districts throughout the United States and Canada, as well as delivering major presentations in Vienna (Austria), Budapest (Hungary), Adelaide (Australia), and San Juan (Puerto Rico). He is in great demand as a speaker at both national and international professional meetings, where his major focus is on preparing *all* students to reason and problem solve in their mathematics classroom, as well as in life.

In 2007, he was given the Great Teacher Award by Temple University. In 2011 he was presented with the Lifetime Achievement Award for Distinguished Service to Mathematics Education by the National Council of Teachers of Mathematics.

Acknowledgement

The authors wish to express their sincere gratitude for the outstanding editing done by Benny Lim Jing Quan and his team. Many thanks to Tan Rok Ting for her excellent guidance throughout the project. The executive editor Rochelle Kronzek as usual has done a wonderful job in initiating the book and moving the process along. Many thanks also to Barbara and Gladys for their patience and support.

Introduction

As today's schools continuously stress the STEM (Science, Technology, Engineering, and Mathematics) subjects, the one area that seems to cause the greatest concern is the teaching and learning of mathematics. It is no secret that some teachers at the lower grades do not relish the task of teaching mathematics, but know they must do it anyway, and therefore do so with a great lack of enthusiasm. Consequently, many schools, particularly at the pre-high school years, employ a talented and seasoned mathematics teacher to serve as a coach for other teachers. There is however, no guarantee that these professionals fully understand the responsibility bestowed upon them.

The *math coach*, as the position is typically known, must stress the ideals of effective instruction, such as

- appropriate classroom management,
- the didactics of the subject matter, including
  - multiple methods to motivate students,
  - appropriate use of technology,
  - incorporating problem solving strategies wherever applicable,
  - and introducing enrichment ideas to make mathematics instruction come alive!

Essentially, teachers should be encouraged to avoid the all-too-common practice in recent years of "teaching to the test".

With a plethora of illustrations and examples, this book is intended to address each of these responsibilities in a meaningful fashion. We begin with a brief look at the various aspects of being a math coach.

1. What is a Coach?

According to most dictionaries, a coach is a person who helps players improve their performance in their chosen field. They serve as coordinators, managers, teachers, and general supporters. They keep current in their field — constantly sifting through all available material to prepare a plan appropriate for their "team" to develop and enhance skills and abilities. The relevance of having such a coach for math education will be apparent after the following point.

2. How Math is Taught Now

In elementary schools, teachers are usually responsible for teaching a wide variety of subjects such as language, history, science, art, music, and of course, mathematics. As mentioned previously, many teachers at this level do not embrace mathematics with interest and thus, teach it in discomfort. It is therefore, delegated as a lesser priority and delivered without any enjoyment or enthusiasm — often also devoid of enrichment or enhancement. This rather negative attitude towards mathematics is subconsciously passed on to students, as the teachers recall it being taught to them. A distaste for mathematics could form among their students that extends throughout and even after their elementary education, having long lasting implications.

Even though mathematics teachers at the middle and high school levels generally have a stronger background in the subject, they are nonetheless challenged as the art of teaching mathematics continues to evolve in technology, problem solving, reasoning and thinking, as well as enrichment of the subject matter. No longer is strict memorization the way mathematics should be taught, and even less so to merely "cover the text." With these changes both in content and pedagogy, even experienced teachers can use support in presenting their subject most effectively to their students. And despite the current trend of rating teachers by student achievement, teachers must be encouraged not to "teach to the test," but rather make their instructional program so effective that favorable student achievements will naturally follow.

3. Enter the Math Coach

It is therefore, no surprise that out of all the subjects, mathematics might require the most support in achieving improvement. The position of math coach has thus assumed a major role in schools due to the needs of teachers. Being a math coach, however, is no simple job; it requires one to enjoy both teaching and learning mathematics, simultaneously possessing a background in the subject far beyond what is taught. The coach must emanate a strong appreciation of the power and beauty of mathematics — this love and enjoyment of mathematics should be so contagious that it easily rubs off onto the students.

The responsibilities of the coach, however, do not simply end with the students. The accomplished math coach (who must already be a certified math teacher) is also a teacher of teachers. Possessing leadership and a wide variety of teaching techniques, the coach imparts these skills upon teachers to bring out the best in both themselves and their students alike.

4. Who Should Read This Book?

Math Coaches: Being the prime intended audience, it is obvious that math coaches should read the book and keep it at ready reference.

Math Teachers: For teachers who work with math coaches, or aspire to be one, this book can improve their understanding of, and ability to transmit, mathematical concepts and skills to their students.

Principals and Senior Administrators: With metrics for performance being heavily dependent on standardized testing, it is not uncommon for math coaches to keep school leaders up to date on the prevailing techniques in math education to further the school's advantage. This book provides an excellent opportunity for such leaders to acquire these techniques themselves and better appreciate the role of math coaches.

The Browser: This guide is full of deep insights into the techniques of math education, so general readers will be exposed to the niceties of learning mathematics in an instructionally effective and motivating environment, while also (hopefully) recognizing the elegance and potential of the subject.

Chapter 1

Math Coaches and Math Teachers

Before delving into the intricacies that accompany the duties of being a math coach, we first broadly define the characteristics, roles, and contributions of the coach and (less so) the teacher, so the reader may better understand their motivations and interactions.

1.1. What Makes a Good Math Coach

Being an external agent, one of the most beneficial contributions a math coach provides is a refreshing "outside" perspective on how the subject is typically taught. Teachers who are often stretched between classes may lack the time and capacity to thoughtfully reflect and improve on their teaching methods; this is where the math coach steps in. The coach gives continuous guidance and advice in both formal and informal settings, updating teachers on new teaching techniques as they are developed. However, not every technique in the arsenal is employed, but instead only the ones tailored to complement both the teachers' personalities and the students' interest and background. These main duties of the math coach aim to address the problems faced in early math education, as discussed in the Introduction (2. How math is taught now). This book, as you will see, strives to equip a math coach (or anyone else, for that matter) with the necessary techniques to carry out these instructional tasks.

But of course, giving good direction is just one facet of being a good math coach. The defining attribute of a math coach should, however, be the embodiment of infinite enthusiasm towards the subject, which will pervade all interaction with both teachers and students. This is not a learned skill nor something this book can provide; it is a passion which is innate.

With great passion there should also be caution. Overconfidence or impractical expectations may come to bite the overly-enthusiastic math coach. You, as a math coach, *do* play an important part in your school's math program, but you are not the ultimate cure-all for any problem related to math teaching. Do trust in the other strategies that the school leaders have devised, and keep in mind that results of your efforts are largely anecdotal in nature and are difficult to observe tangibly. Positive results will come, but slowly; don't expect major changes overnight.

1.2. General Responsibilities of a Math Coach

The first change you bring about as a math coach is surprisingly non-technical in nature. You start by improving the overall attitude and outlook of both teachers (mainly) and students towards teaching and learning mathematics. Many elementary school teachers are quite literally afraid of teaching mathematics, so much that the term "math phobia" was formally defined in the 1950's. Consequently, negative attitudes and a lack of competence in math instruction propagate down to successive generations. Yet many experienced teachers stubbornly say, "I've been a success," or "It has worked for me!" Their idea that "We have been successful in what we do and have done for many years. Why change now?" is challenged by the fact that math education has evolved to focus on reasoning, thinking, and problem solving, spurred by great change to teaching techniques due to technological advances. This adverse persistence extends beyond attitude, to teachers' understanding of "finishing" the curriculum — the learning journey is considered complete so long as every page of a textbook is covered, while it should only serve as a reference. It is, thus, the duty of a math coach to sway these beliefs and preconceptions for the better.

The key to great math education in the elementary school level is the delivery of lessons by teachers, as younger students are more susceptible to influence from mentors than their own opinions. Math coaches build on this significance of teachers' direction by enlightening the teachers themselves on how technical knowledge translates to relatable real-world applications, enhancing their presentation to the students. This method of enriching lesson delivery is effective not only in elementary classes where the focus comprises whole number operations, but also at middle to high school levels where they start to grasp geometry and algebra.

Before going into instructional intricacies, it is worth briefly noting the social and personal interactions between yourself, the math coach, and the teachers you assist. A degree of mutual trust and respect must be established to maintain a healthy professional relationship. Try to build a feeling that you are there to help improve teaching mathematics and not to "spy" on the teachers or evaluate them. A good starting point is to always make an appointment or obtain permission before dropping in on a class otherwise unannounced; and if the teacher has a hand in inviting the coach, the atmosphere will be more relaxed and easy going.

1.3. Specific Coaching Strategies

Another way to keep the classroom environment relaxed and refreshing is to implement a simple change in the seating arrangement. Most traditional classrooms still have desks arranged in a series of rows and columns. This inadvertently encourages the teacher to stay in the front of the room while lecturing. But what about that youngster at the back? Will he pay attention or just drift off to sleep? How can the teacher best monitor this? The coach might simply suggest a U-shaped array of desks so that no student sits behind another, and it becomes much easier for the teacher to monitor each student in this new-feeling classroom.

Apart from contributions within the classroom, there are many parts of a coach's job that can take place outside the regular curriculum. One good example would be to promote competition in math. If the schools in your area already have a sort of math league or competition established, you can help form a math team in your school for those students who would enjoy competing in math. The team would meet for practice during a club period or after the school day, and be shown how to solve unusual math problems as well as learn new problem-solving techniques that can transfer beyond competition to the math class and to everyday life. Training the students in concepts beyond the ordinary syllabus equips them with a strong advantage to compete in the league. However, if a league does not already exist, you might consider contacting the coaches/teachers in nearby schools to determine if enough interest exists to form such a league. The hardest part of your job here would be to find someone to write the actual problems for the inter-school contest that would take place, say, once a month. The league format could go as such: All schools in the math league would meet at rotating school locations and score points for successfully solving problems. The league coordinator would receive the scores from the team coaches and keep a running account. At the end of the year/term, the top teams would each receive a trophy or other prize.

In addition to establishing a league, the math coach could also organize an in-school "Problem of the Week" contest which could be run in this way: Every Monday morning before the school day begins, the coach would post a problem for the students to try to solve. Each of these problems should be challenging, but manageable by the average to above average students, and solutions should be submitted to the math coach on an ongoing basis. On Friday morning, the coach would commend and post the names of the students who had correctly solved the problem. Exemplary solutions would also be recognized and posted, such that teachers will be encouraged to review the "Problem of the Week" in their classes to share its enriching nature, in turn motivating more students to participate.

However, not all the math coach's contributions need be pertinent to hard skills (solving math problems, teaching new methods, etc.). The coach should also advise on the softer, emotional aspects as well. In the way mathematics is taught now, students are encouraged to memorize arithmetic rules and methods routinely to solve "problems" (or "exercises" as they should be called). This archaic teaching technique has led to very little original thinking from the students, causing them to develop a purely mechanical approach towards studying mathematics. How can you, as the math coach, help your teachers rid themselves of this non-constructive attitude? You can first attempt to demonstrate the beauty of mathematics to teachers. For example, ask your teachers which is greater, (b) or (–b)? Encourage the teachers to ask questions about the problem and discuss their questions. In fact, welcome them to ask questions at any time and about any problems they may have, which you should always take seriously and attempt to answer. You might even locate an online refresher course in mathematics that the teachers examine at their leisure to reignite their interest in mathematics which will surely propagate to their students.

Many teachers also erroneously feel that both their students and themselves will be evaluated based on test results. But nothing could be farther from the truth. The test should never be the sole reason for teaching a new skill or topic, and we might add that this feeling leads to "teaching to the test", which is counterproductive to good teaching practice. It is hence, the coach's responsibility to dissuade teachers from thinking this way, but instead steer them towards teaching mathematics purely for its beauty and power.

1.4. Further Enrichment Outside Curriculum

We have mentioned how students should be encouraged to learn math outside regular curriculum. But there also exist other social, more transferable skills we may teach. Many people in industry are looking for a different kind of student who graduates in today's world — they want someone who is adept at communicating ideas. The coach plays an important role in nurturing communication. As we said earlier in the chapter, changing the seating arrangements can make a big difference. As simple an act as reseating children in pairs helps them learn how to communicate new and different ideas and practice communication techniques. This especially offers an excellent opportunity for teaching problem solving via strategy games. For example, most students are familiar with the game of Tic-Tac-Toe, where three in a straight line wins. But what if we change the goal of the game? Now let's make it three in a row *loses*. The pairs of students can easily discuss what this change means and how to win at this new problem-solving-strategy game. What does "win" mean? Would it be advantageous to go first or second? And why? As the pairs of student talk to one another they try to discover what would constitute a winning strategy, and at the same time reinforce their communication skills.

Another skill required to succeed in today's world doesn't depend so much on what specific math skills have been learned (although these still are very important). Rather, it is *how* to learn, to adapt to new ideas that employers look for when hiring new employees. Problem solving and reasoning are very much in demand. As problems arise, do the students know how to react? Can they develop a method of attack? The coach helps teachers convey these practical skills to their students, as we will see later in this book.

Chapter 2

Coaching Teachers to Effectively Motivate Their Classes

To impart any sort of knowledge or skill effectively, teachers must be good motivators to begin with. They must constantly evaluate and determine the best teaching strategies in the classroom such that students will be most attentive while enjoying the learning process. Achieving this largely depends on the delivery of the content rather that the content itself. If students are merely told at the onset all the information about a topic, they will inevitably become bored, lose interest, and ultimately "tune out." Thus, the teacher must constantly try to promote an exciting learning atmosphere in the classroom by encouraging mental stimulation throughout the lesson. Such stimulation can come with regularly offering moderate mental challenges (in either small or large group settings). In this chapter, we will present refined techniques and strategies that are designed to motivate and stimulate student thinking, while demonstrating more effective ways to teach.

2.1. What Is Motivation?

How to motivate students to learn is the crux of a teacher's concerns, for if students can be made to be delightfully receptive learners, then the rest of the teaching process becomes significantly easier and profoundly more effective.

Naturally, when thinking of how to "make a student want to learn" what you are about to teach, certain *extrinsic* methods of motivation may come to mind. These may include token economic rewards for good performance, peer acceptance of good performance, avoidance of "punishment" by performing well, praise for good work, and so on. Students' upbringing and environment have much to do with their adaptation of commonly accepted extrinsic motivators. However, many students demonstrate *intrinsic* goals in their desire to understand a topic or concept (task-related), to outperform others (ego-related), or to impress others (social-related). The last goal straddles the fence between being an intrinsic and an extrinsic goal. In a more structured form, ***intrinsic motivators*** tend to conform to the following basic types:

The Learner Wants to Develop Competencies. Students are often much more eager to do a challenging problem than one that is routine. It is not uncommon to see students beginning their homework assignment with the most difficult problems presented in the form of a "challenge for experts," even if the time spent on this prevents them from completing their routine work.

The Learner Is Curious about Novel Events and Activities. It is a natural human trait to seek out unusual situations or challenges that can be conquered by existing skills and knowledge and, thereby, provide a feeling of competence. When the learner's curiosity about unusual stimuli is piqued, it becomes a form of motivation.

The Learner Has a Need to Feel Autonomous. The desire to act on something as a result of one's own volition is often a motivating factor in the general learning process. To determine for oneself what is to be learned, rather than to feel learning is being done to satisfy someone else or to get some sort of extrinsic reward is another basic human need.

The Learner Reacts with Some Internalized Social Values. Not to be overlooked when trying to simplify (and catalog) human needs and motives is the notion that all learners have certain moral values that have been internalized through years of social reinforcement — most often in the home environment. For example, if a parent constantly tells a child that hard work is good, then that value is manifested in the child and becomes a part of the motives that drive the child. The teacher's task is to understand the basic motives already present in the learners and to capitalize on them. The teacher can then capitalize on this knowledge of the students' motives to maximize the effectiveness of the teaching process. Often, this process can result in some rather artificial situations, contrived specifically to exploit a learner's motives in order to generate a genuine interest in a topic. This is eminently fair and highly desirable!

With these basic concepts in mind, we seek to explore how they can be used to motivate mathematics instruction. Naturally these specific techniques should be expanded, embellished, adapted to the teacher's personality, and, above all, made appropriate for the learner's level of ability and environment.

2.2. What are Motivational Techniques?

Planning a motivational activity requires not only creativity and imagination, but also careful consideration of students' needs and interests, which naturally vary.

We sometimes rely on certain aspects of mathematics as being self-motivating. Such as in geometry, because of its appealing visual nature, teachers tend to believe that this subject needs no additional effort to motivate students. This is unfortunately not always the case, since much of the course deals with proving theorems, and then applying them to artificial problems. Also, the way these theorems are presented and proven tends to revert to an automatic process as teachers feel that such a technique will allow for the material to be more easily transmitted to their students. Nonetheless, there are always students who will be enthralled by the visual aspects of geometry. But even so, a teacher should address the total population, of which a majority may not share this sentiment. Furthermore, a good teacher must focus attention on the less-interested students, who may be less enchanted by the alluring nature of geometry, and therefore, need a good motivational activity at the beginning of every lesson.

To motivate students is to channel their interests to the specific topic to be presented in a particular lesson. Let us now consider some techniques that you and your teachers can use to motivate students in their mathematics instruction. We will present ten techniques that can be used along with some illustrative examples at various levels. These examples are presented merely to better explain the techniques presented. However, a good coach will seek appropriate applications for the intended population.

2.3. Ten Techniques for Motivating Students

2.3.1. *Indicate a void in students' knowledge*

It is not uncommon for people to want to complete either a collection, such as a set of stamps, or understanding of a subject. Students usually have a natural desire to complete their knowledge of a topic, especially if in a diplomatic way that notion is made clear to them. This motivational technique involves making students aware of a void in their knowledge and capitalizes on their desire to learn more. For instance, you may present a few simple exercises involving familiar situations followed by exercises involving unfamiliar situations on the same topic. Or you may mention (or demonstrate) to your class how the topic to be presented will complete their knowledge about a particular part of mathematics. The more dramatically you do this, the more effective the motivation. Often, *guiding* students to discover this void in their knowledge is effective. Following are some examples of how this technique may be used.

Example 2.1: If you are introducing the notion of a general angle in second-year algebra you might provide students with the following challenge, namely, to find the value of each of the following without the aid of a scientific calculator:

1. sin 30°
2. cos 60°
3. cos 120°

Students familiar with the relationship of the sides of a 30-60-90 triangle ought to be able to answer the first two questions easily. The third question will cause students some discomfort, since students are unfamiliar with trigonometric functions of angles whose measure is greater than 90°. You should now have students realize that there is a void in their knowledge, hence will be motivated to fill that void, which is the topic of the ensuing lesson.

Example 2.2: You can cleverly indicate a void in the students' knowledge by introducing the topic of finding an angle with its vertex **outside** of a given circle and whose rays intersect the circle. We will, of course, assume that the students have already become familiar with the relationship of an angle formed by two intersecting chords of a circle, with the point of intersection either **in the circle** or **on the circle**. The motivational activity would be to have the students find the value of x in each of the circles in Figure 2.1.

The students should feel comfortable that they could do the first two problems. However, when the vertex is now outside the circle, as it is in the third problem, the students will feel challenged and have an inner desire to know how to find the measurement of such an angle.

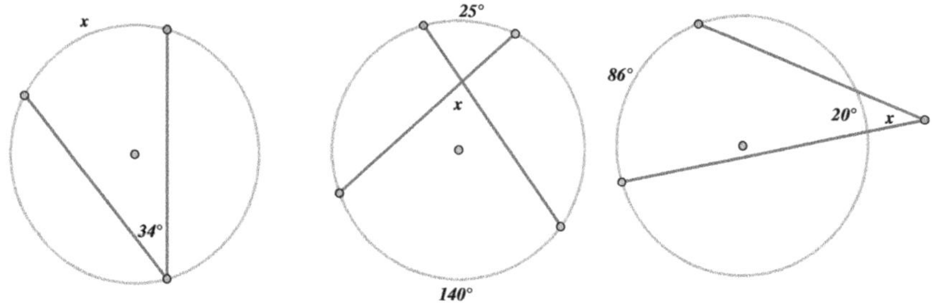

Figure 2.1

Example 2.3: There are times when a simple question about a topic which students already know could also be motivating in that they are challenging them within the topic. Consider the following, where some interesting questions that can be used to motivate a classroom discussion are offered:

1. Which is greater, a or $-a$?
2. Which is the largest, π, 3.14, or $\frac{22}{7}$?
3. For what values of x is $x-3$ greater than zero?
4. Is $\sqrt{x} > x$? Is that true for all values of x?
5. What is the result of squaring $\sqrt{\frac{1}{2}}$?

2.3.2. *Showing a sequential achievement*

Much of what we teach in mathematics is done sequentially based on prior knowledge. However, teachers need to demonstrate the logical sequence of many of the topics presented.

Example 2.4: When presenting the properties of the various quadrilaterals in the study of geometry, point out the relationship between various parallelograms — the square, the rectangle, the rhombus in sequential fashion. This can be quite motivating when done with that aspect in mind. A diagram showing the relationship as it develops can motivate students to go from one quadrilateral to another. A chart might be developed, with its intended purpose clearly in focus.

2.3.3. *Enabling students to discover a pattern*

There is hardly any subject other than mathematics that can give a special kind of structure to pupils. Most people have a natural desire for order, perhaps because it offers some sort of security, or something one can refer to in the process of problem solving.

Math can be comforting and reassuring, as new concepts can be made accessible from previous knowledge that has been at a student's disposal. If the teacher manages to cleverly set up a situation that leads students to discovering a pattern, it can often be quite motivating. They then feel great pleasure in finding and then owning an idea.

Naturally, the teacher can be an important supporter in this process, as he or she may carefully draw a student's attention towards a somewhat hidden pattern. The teacher's guidance should be so skillfully inconspicuous that students never lack the feeling of having discovered it for themselves. Such realization of one's own ability is essential for the improvement of understanding and cognition.

Knowing different ways to resolve problems, and having discovered patterns on one's own, are normally very long-lasting experiences, which can be used in future contexts successfully. As a teacher, one should keep this notion in mind, when considering merely writing a new formula on the board without allowing at least some modicum of student discovery. The reason for such an oversight might be a perceived lack of time and the focus on being able to promptly use this formula to deal with certain ensuing exercises. Often these formulas can be deduced in an easy and elegant way, using means that are at the student's disposal, notably discovering a pattern.

Example 2.5: There are times when students begin to lose faith in the usefulness of algebra. This stems from often repeated exercises that are used to firm up students' ability to do algebraic manipulations. To motivate a lesson that would expose the background for some unusual patterns can certainly be a form of motivation. You may show students why a number is divisible by 3 or 9, when and only when, the sum of its digits is divisible by 3 or 9, respectively. There are many other rules for divisibility, which can be very nicely justified through algebra. Presenting such patterns at the start of a session, where a review of algebraic skills is the topic of the day would be a fine form of motivating the rest of the session.

Many students, and adults as well, have a distinct fear of mathematics. The only chance to fight against this widespread phobia is to awaken a deeper insight into the subject. But this can only be successful if instruction not only imparts a body of rules and concepts, which can be recalled from memory as needed, but also shows ways and structural approaches to mathematical tasks and exercises. This is nicely done through the recognition of patterns, which in proper measure can be a most effective motivating device though not always applicable, as we shall show in this chapter with a variety of topics from the secondary curriculum.

We encounter patterns every day and they pervade all aspects of our life. Our ability to recognize patterns and to create them is how we deal with the world and as a result, many mathematicians consider patterns as either the basis for all mathematics or as a unifying theme. Revealing hidden patterns is what mathematicians do! The search for patterns inspire the growth of mathematics and are often applied to explaining natural phenomena. Essentially, students go through three stages when discovering patterns.

1. They first collect their data and try to organize it.
2. They then attempt to classify the information they have collected.
3. Once they have put their data into some kind of order, they try to find a pattern to explain the data.

The more practice and experience the students have with finding and recognizing patterns, the more adept they become at recognizing patterns from data. Although problems that involve finding patterns are numerous, you must search for those that might interest your students. Some patterns may not be mathematical in content yet provide a great deal of motivation for the class.

Example 2.6: You may ask your students to find the pattern rule for the following sequence:

$$8, 5, 9, 1, 7, 3.$$

Students may groan when they find the pattern rule, namely, the numbers are listed in order of the first letter of their spelling (**E**ight, **F**ive, **N**ine, **O**ne, **S**even, **T**hree). They may feel similarly frustrated in discovering the pattern in the following sequence:

$$1, 8, 3, 6, 5, 4, 7, 2, 9.$$

Here the sequence consists of two interspersed sequences of the ascending odd numbers (1, 3, 5, 7, 9) and the descending even numbers (8, 6, 4, 2).

Setting up a situation that leads students to "discovering" a pattern can often be quite motivating as we have said before: students take pleasure in finding and then "owning" the idea. We have been working with patterns since we were mere toddlers — a very young child separating marbles: "the big ones go here; the small ones go there" is a form of pattern development. As we mature, we observe and recognize more and more complex patterns. Observing a pattern and continuing to apply it can often lead directly to the resolution of a particular problem situation. Presenting activities that lead to pattern recognition is an excellent motivator for students in middle and senior high school grades.

Students who have just begun to understand the nature of positive integer exponents will probably respond to the question,

"What does 5^n mean?" with a response such as "the product of n factors of 5." When asked what the nature of n is, they will probably say it is a positive integer. This motivator will encourage them to consider the non-positive integers: 0 and the negative integers.

Example 2.7: You might show that these definitions enable an observed pattern to continue. Consider the following:

$$3^4 = 81$$

$$3^3 = 27$$

$$3^2 = 9$$

$$3^1 = 3$$

Then continuing this pattern of dividing the result by 3 while decreasing the exponent by 1, we obtain:

$$3^0 = 1$$

$$3^{-1} = \frac{1}{3}$$

$$3^{-2} = \frac{1}{9}$$

$$3^{-3} = \frac{1}{27}$$

Students should use this pattern to motivate further examination of these negative exponents. They may also look at the issue by considering $\frac{x^5}{x^5}$ (where $x \neq 0$), which equals $x^{5-5} = x^0$. Therefore, $x^0 = 1$. However, the following is a meaningless statement: "x used as a factor 0 times is 1." To be consistent with the rules of exponents, we *define* $x^0 = 1$; then it has meaning.

In a similar way, a student cannot verbally explain what x^{-4} means. What would it mean to have "x used as a factor −4 times"? Using the rules of exponents, we can establish a meaning for negative exponents. Consider $\frac{x^5}{x^8} = \frac{1}{x^3}$. By our rules of operations with exponents we find that $\frac{x^5}{x^8} = x^{5-8} = x^{-3}$. Therefore, it would be nice for $x^{-3} = \frac{1}{x^3}$, and so we can *define* it this way and our system remains consistent.

Once we arrive at $\frac{x^k}{x^k} = x^{k-k} = x^0$, we can now consider what value 0^0 should have. Using the same idea, we get $0^0 = 0^{k-k} = \frac{0^k}{0^k}$, which is meaningless, since division by $0^k = 0$ is undefined. Similarly, we cannot define 0^{-k}, as this yields $\frac{1}{0^k}$, which is undefined. Students can thus conclude that the base cannot be 0 when the exponent is 0 or negative.

Therefore, the definitions $x^0 = 1$, given $x \neq 0$, and $x^{-k} = \frac{1}{x^k}$, given $x \neq 0$, become meaningful. The use of patterns motivates the students to keep exploring in this fashion to get a more solid understanding of exponents.

2.3.4. *Present a challenge*

Great care must be taken when crafting a challenge for students. While the right challenge will elicit boundless enthusiasm, an ill-fitted one may lead to frustration and disinterest. The right challenge is short and not too complex, engrossing, but not overly so that it distracts from the intended lesson, and within the grasp of the students which ultimately leads into the desired lesson. Immaculate judgment must be applied by the teacher here to develop the most effective challenge.

Example 2.8: Have students determine the perimeter of triangle *ABC* in Figure 2.2, where the triangle is formed by three tangents to the circle. They are then given that the length of $AR = 36$.

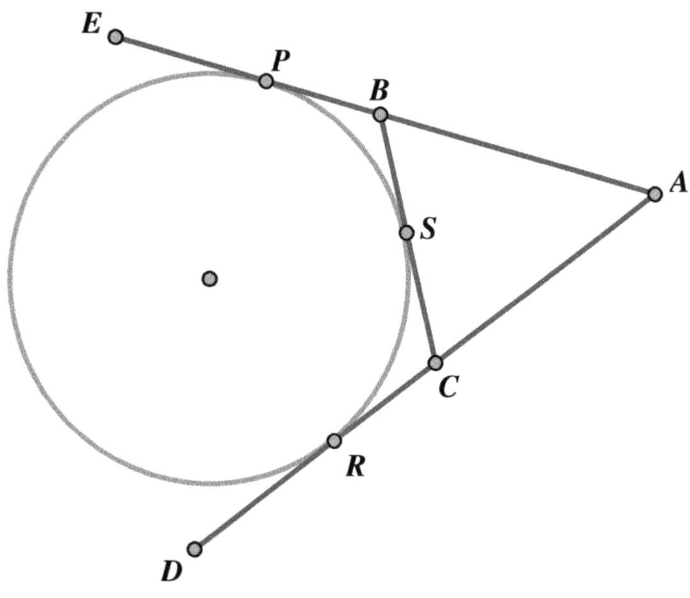

Figure 2.2

This can lead nicely into the lesson on tangents to the same circle from an external point have the same length. They will realize that $AR = AP$, and since $BP = BS$, and $CS = CR$, the perimeter of the triangle is simply 72.

Example 2.9: Here is another challenge that is a fine motivational beginning to a lesson on the sum of a geometric series. Ask a class which they would rather have:

1. \$100,000 per day for 31 days, or
2. 1 cent the first day, 2 cents the second day, 4 cents the third day, 8 cents the fourth day, 16 cents the fifth day, and so on for the 31 days.

Experience shows that most students will most likely select the first option, since that results in \$3,100,000. However, they may be motivated to determine the sum of the second option. Astonishingly, they will find (with your help) that the sum is \$21,474,836.47.

An effective way of coming up with challenges such as these is to have one teacher prepare a draft and presenting it to other fellow teachers. This may be done with reference to textbooks or purely with the teacher's creativity. Discuss whether this motivational beginning would gain student interest. Why or why not?

Teachers may use counterexamples which provides an interesting motivation for a lesson. This technique might be difficult to employ at first hence discouraging many teachers from using it. You, as the coach, may prepare a worksheet similar to the one that follows as a guide. This topic provides an excellent opportunity for discovery and for the teachers to discuss the use of quantifiers such as "all", "some", and "none".

Example 2.10:

Worksheet on Counterexamples

Find an example and a counterexample for each of the following.

1. For all real numbers x and y, $\sqrt{x}+\sqrt{y}=\sqrt{x+y}$.
2. Rectangles with equal perimeters have equal areas.
3. Rectangles with equal areas have equal perimeters.
4. Triangles with equal altitudes have equal areas.
5. For all real numbers m and p, $\frac{1}{m}+\frac{1}{p}=\frac{1}{m+p}$.

6. Negative numbers are greater than zero.
7. Equilateral triangles are congruent.
8. For all real values of x, $-x < x$.

Example 2.11: Another interesting challenge is to tell the class to write any six-digit number of the form *abcabc* where a, b, and c are positive integers. Then tell your students to verify one of the following statements:

1. Your number has a factor of 7.
2. Your number is divisible by 13.
3. Your number has a factor of 91.

It can be shown that the original number *abcabc* can be written as:

$100{,}000a + 10{,}000b + 1000c + 100a + 10b + c$, which yields $1001(100a + 10b + c)$. Yet, we know that the number $1001 = 7 \times 11 \times 13$. Regardless of the values of a or b or c, the number *abcabc* is divisible by (or has a factor of) 7, 11, and 13, or for that matter $7 \times 13 = 91$.

2.3.5. *Indicate the usefulness of a topic*

Like the challenges previously discussed succinct but interesting practical applications of math may be introduced at the beginning of a lesson. These serve the same purpose as the challenges before – to elicit interest before a particular topic is taught. To best ensure this, teachers should tailor their choice of examples, such that students will have prior knowledge of the concept being applied.

Example 2.12: There are certain times when the teacher can present a situation that may not have been part of the students' experience, nonetheless make it relevant. Suppose one wants to erect a flag pole that would be perpendicular to the level ground. How might that be most easily done? This question could lead into the property of a line perpendicular to a plane. Here, the teacher needs to guide the students to realize that in order for a line to be perpendicular to a plane, it needs to be perpendicular to each of two intersecting lines in the plane at the foot of the perpendicular.

Example 2.13: This is a useful technique in geometry that can lead nicely into the lesson on the property of the segments of two intersecting chords and the circle. Consider a portion of a broken plate where one has to find the original size (i.e., the diameter) of the plate so as to order a replacement plate. All that is left of the plate is a small portion of the plate shown in Figure 2.3 as the region formed by the segment *MSN* (that is, the region between arc *MSN* and line segment *MN*) of the circle *O*.

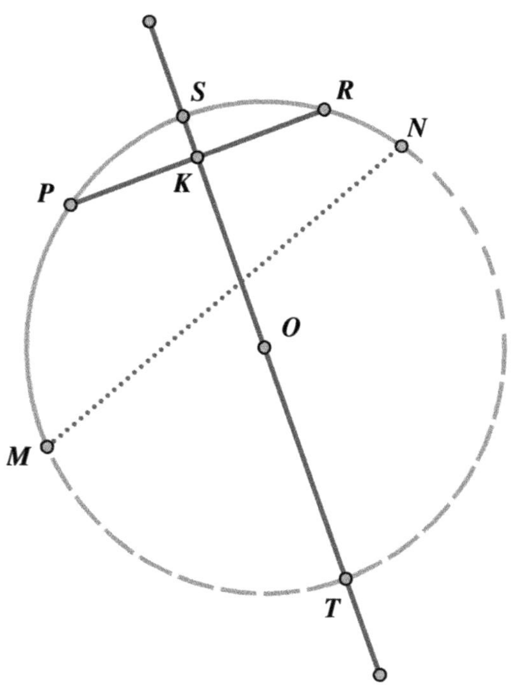

Figure 2.3

The only significant length that the student would have available would be that of a chord, such as *PR*. By drawing the perpendicular bisector of the chord *PR*, the students can be shown that the product of the segments *PK* and *RK* is equal to the product of the segments *SK* and *TK*. Since the length of *SK* can be measured, the length of *KT* can be found by the relationship that will be developed in this lesson, namely, that the product of the segments of one chord is equal to the product is segments of the other chord. Once they have found the length of *KT*, they will have been able to calculate the diameter, *SKT*, of the circle.

2.3.6. *Use recreational mathematics*

Recreational motivation consists of puzzles, games, paradoxes, or facilities. In addition to being selected for their specific motivational gain, these devices must once again be brief and simple. A student should realize the "recreation" without much effort in order for this technique to be effective. Particular caution must also be taken to select a motivational activity in this style which will not detract from the ensuing lesson but rather enhance it through motivation.

Example 2.14: At a point at which a teacher wants to demonstrate the importance of knowing that division by zero is not acceptable, the teacher can produce a "proof" that $1 = 2$; clearly an absurdity! The teacher should have the class agree to begin with

$$a = b,$$

and then multiply both sides of this equation by a to get

$$a^2 = ab.$$

Then by subtracting b^2 from both sides of the equation we get

$$a^2 - b^2 = ab - b^2.$$

Factoring both sides of the equation gives us

$$(a + b)(a - b) = b(a - b),$$

and dividing throughout by $(a - b)$ yields

$$a + b = b,$$

but since $a = b$, we could write this as

$$2b = b,$$

which is the same as

$$2 = 1.$$

At this point, we have had some recreation, which probably has led to a bit of frustration on the part of the students. Now they can be led to believe that the error occurred in both sides of the equation when divided by $(a - b)$, which in fact had a value of zero, since $a = b$ as defined in the beginning.

Example 2.15: Consider the following situation: Select any three-digit number in which the hundreds digit and the units digit are not equal. Then have them write the number of digits in the reverse order of the selected number. The teacher then should tell them to subtract the smaller number from the larger. Once again, they are to take the difference that was obtained and reverse its digits and add a new number to the previous number. To everyone's astonishment (those who did it correctly) they will end up with the number 1089.

To illustrate, suppose a student selected the number 934. Her computation would appear as:

934	(original number)
439	(reversed digits)
495	(difference)
594	(reversed digits)
1089	(sum of the last two numbers)

When students compare their results, they will be amazed to discover the uniformity in their answers. At this point, they should be quite eager to find out *why* they all came up with the same result.

2.3.7. *Tell a pertinent story*

A story of a historical event or of a contrived situation can be quite motivating indeed. But all too often teachers rush through the story as they already know it well and are eager to get into the "meat" of the lesson. Such a hurried presentation minimizes the potential effectiveness the story may have as a motivational device. Thus, a carefully prepared method of presentation of a story is almost as important as its actual content.

Example 2.16: One such story that can be used very nicely to introduce the subject of finding the sum of an arithmetic sequence is to tell about the young Carl Friedrich Gauss, one of the greatest mathematicians in our history. When he was about 10 years old, his teacher wanted to keep the class busy and asked that they all add up the numbers from 1 to 100 on their slate boards. No sooner had the teacher given this assignment that the young Gauss raised his hand indicating that he already had the answer. The teacher ignored this and told him to be quiet and keep working. One-half hour later, the teacher asked the students to give the answer to the question and the only one who had the answer correct was Gauss, who got the answer within one minute of having been given the assignment. The teacher was amazed at how quickly and accurately Gauss got the answer to which Gauss explained that rather than adding the numbers sequentially: $1 + 2 + 3 + 4 + \ldots + 97 + 98 + 99 + 100$, he added the numbers in pairs as follows:

$$\begin{aligned}
1 + 100 &= 101 \\
2 + 99 &= 101 \\
3 + 98 &= 101 \\
&\vdots \\
48 + 53 &= 101 \\
49 + 52 &= 101 \\
50 + 51 &= 101
\end{aligned}$$

realizing that he had 50 pairs of 101, the answer was very simply $50 \times 101 = 5{,}050$. This scheme can then be used to develop the formula for the sum of an arithmetic series.

Example 2.17: There are lots of other stories that can be used to introduce topics in mathematics. One other such example would be how Eratosthenes measured the circumference of the earth by using shadows at different places. This story can easily be found and verified with a short research.

An endless number of stories apart from these two, historical and otherwise, can be used for motivational purposes. Here are just a few of many books available to provide stories that can be used to generate interest at the start of the lesson. As with the other methods, stories selected must be directly relevant to the topic presented, appropriate the level of the class, and brief enough that they do not dominate the lesson.

2.3.7.1. *Ancient measuring devices*

1. Kline, Morris. *Mathematics: A Cultural Approach.* Reading, MA: Addison-Wesley, 1962.
2. Martzloff, Jean-Claude, *The History of Chinese Mathematics.* New York: Springer-Verlag, 1997.
3. Polya, George. *Mathematical Methods in Science,* Washington, DC: Mathematical Association of America, 1977.

2.3.7.2. *Major breakthroughs in mathematics*

1. Bunt, L., P. Jones, and J. Bedient. *The Historical Roots of Elementary Mathematics.* Englewood Cliffs, NJ: Prentice-Hall, 1976.
2. Smith, David E. *A Source Book in Mathematics.* New York: McGraw-Hill, 1929.
3. Newman, James R. *The World of Mathematics* (4 vols.). New York: Simon & Schuster, 1956.
4. Struik, D. J. (Ed.). *A Source Book in Mathematics, 1200–1800* Princeton, NJ: Princeton University Press, 1986.

2.3.7.3. *Pertinent biographical notes*

1. Bell, E. T. *Men of Mathematics.* New York: Simon & Schuster, 1937.
2. *Biographical Dictionary of Mathematicians,* Volumes 1–4. New York: Charles Scribner's Sons, 1991.
3. Coolidge, Julian L. *The Mathematics of Great Amateurs.* New York: Dover, 1963.
4. Gindikin, S. G. *Tales of Physicists and Mathematicians.* Boston: Birkhaüser, 1988.
5. Perl, Teri. *Math Equals Biographies of Women Mathematicians and Related Activities.* Menlo Park, CA: Addison-Wesley, 1978.
6. Posamentier, Alfred S. and Christian Spreitzer. *Math Makers: The Lives and Works of 50 Famous,* 2019.
7. Schmalz, Rosemary. *Out of the Mouths of Mathematicians.* Washington, DC: Mathematical Association of America, 1993.
8. Turnbull, Herbert W. *The Great Mathematicians.* New York: New York University Press, 1961.

2.3.7.4. *Pertinent anecdotes*

1. Aaboe, Asger. *Episodes from the Early History of Mathematics.* Washington, DC: Mathematical Association of America, 1964.

2. Dunham, William. *Journey Through Genius: The Great Theorems of Mathematics.* New York: John Wiley and Sons, 1990.
3. Eves, Howard. *In Mathematical Circles* (2 vols.). Boston: Prindle Weber and Schmidt, 1969.
4. Eves, Howard. *Mathematical Circles Revisited.* Boston: Prindle Weber and Schmidt, 1971.
5. Kaplan, Robert. *The Nothing That Is, A Natural History of Zero.* New York: Oxford University Press, 1999.
6. Katz, Victor J. *Using History to Teach Mathematics.* Washington, DC: Mathematical Association of America, 2000.
7. Posamentier, Alfred S. and Ingmar Lehmann, *Magnificent Mistakes in Mathematics*, Amherst, NY: Prometheus Books, 2013.

2.3.8. *Get students actively involved in justifying mathematical curiosities*

While telling an interesting mathematical story may passively spark a student's curiosity, encouraging them to justify a pertinent mathematical phenomenon is a more effective and active approach. Once again, the students should be comfortably familiar with the mathematical result before you "challenge" them to justify it. Although this could consume more time than may be normally allotted for a motivational activity, to proceed with a justification before sufficient exposure has been achieved would be counterproductive.

Example 2.18: Ask your students to draw any randomly created quadrilateral and join the midpoints of consecutive sides. They will all find that the resulting inside quadrilateral is a parallelogram. However, why this is a parallelogram can be best shown by the geometry theorem to be introduced in this lesson, which is that the line joining the midpoints of two sides of a triangle is parallel to the third side and half its length.

Example 2.19: The coach may procure a *pantograph*, which is a very popular toy used to copy drawings, making them larger or smaller than the original based on simple geometric principles involving proportionality. Verifying the mathematical mechanics involved in the toy's operation naturally leads to a lesson on the similarity of triangles.

2.3.9. *Entice the class with some amazements*

When introducing a topic that could be either done in a typical academic fashion or to introduce it through a mathematical amazement, the latter would have far more effect in motivating the class.

Example 2.20: When introducing the topic of probability, the famous birthday problem would be one way of indicating the power of understanding probability. Tell the class of about 30 students that the chance of there being two students having the same birthdate is around 70%. When there are 366 possible birthdates, this is a truly an amazing result and can be justified very easily using the most elementary probability. If they wish, they could take the first 30 United States presidents and find that there are two who share the same birthdate, namely the 11th President James K Polk, born on November 2, 1795 and the 29th President Warren G Harding born on November 2, 1865. To bring this to even further, counterintuitive results occurs when you indicate that when there are 55 people in a room the chance of two having the same birthdate is 99%!

While we are still on the topic of probability, note that teaching this topic often presents difficulty in motivating the children as the use of dice-tossing or coin-tossing is rarely interesting. Here are more examples of motivational activities that amaze children and challenge their curiosity.

Example 2.21: Have the students write the numbers from 1 through 20 on a blank sheet of paper. Tell them this will be a true-false test, but there are no questions. They are to guess whether an answer will be true (T) or false (F) by labelling each number T or F. Then reveal the "Answer Key" which was formed by tossing a coin, where heads meant a true answer, tails meant a false one.

1. T	6. F	11. F	16. F
2. T	7. T	12. T	17. T
3. T	8. F	13. F	18. F
4. F	9. F	14. F	19. T
5. T	10. T	15. T	20. T

The students should wonder why there are not 10 Ts and 10 Fs. Because after all, their knowledge of probability would expect a 50% chance of obtaining a T and 50% of obtaining an F. This can lead to a discussion of theoretical probability versus actual probability. If you now combine all their results, the percent should move even closer to 50%–50%.

Example 2.22: We take a bag and put in one red checker (R1) and two black checkers (B1) and (B2). You win if you randomly draw two checkers of the same color. Here are the possible results:

R1 B1   R1 B2   B1 B2

You have a one in three chance of winning.

You may now add one checker of either color. Which should you choose, red or black? Most students will suggest adding a second red checker. However here are the results now;

R1 R2	R1 B1	R1 B2
R2 B1	R2 B2	**B1 B2**

The results are 2 out of 6 or 1 out of 3 — same as before.

Let's add a third black checker. Here are the results:

R1 B1	R1 B2	R1 B3
B1 B2	**B1 B3**	**B2 B3**

The results are 3 out of 6 or 50%. A fair game!

This activity usually mystifies teachers as well as their students.

Apart from these amazements involving probability, there are lots of sources where teachers can access these unintuitive mathematical phenomena. One such is "*Mathematical Amazements and Surprises*" by A. S. Posamentier and I. Lehmann (Prometheus Books, 2009).

2.3.10. *Use teacher-made or commercially prepared materials*

Finally, motivation can be achieved by presenting the class with concrete material of an unusual nature. This may include teacher-made materials, such as models of geometric shapes, geo strips, specifically prepared PowerPoint slides, or practical "tools" that illustrate a specific geometric principle. Some fine commercially prepared materials are also available, ranging from geometric models to every-day tools that are based on geometric principles. Materials selected should be reviewed carefully and their presentation carefully planned so as to motivate students for the lesson and not to detract attention from it. Although this method of planning the motivation for a lesson is time-consuming, challenging, and at times difficult, the rewards are immeasurable. The higher degree of student learning resulting from a well-planned and well-executed activity will make this additional work worthwhile.

For example, geometric drawing programs such as Geometer's Sketchpad or Geogebra not only enhance the instructional program but they are also highly motivating for youngsters who might want to experiment with geometric phenomena. When dealing with the topic of similarity a pantograph could also be a useful instrument to exhibit relationships. It also might be motivating for youngsters to see examples from newspaper advertisements that can be explained mathematically and at the same time demonstrate concepts presented in the curriculum. For example, there are often advertisements when a store is holding a holiday sale on all of its items, where they indicate a 20% discount on their already 10% discounted prices. Most consumers think that this is an actual discount of 30%. Well, it is not! This opens the door to a discussion of successive discounts. Here the 20% plus the 10% discounts is actually equivalent to a discount of 28%. In short, the math coach should always be on the lookout for materials that can enhance instruction.

2.4. Summary

In conclusion, remembering a few general rules for using these motivational techniques will make them more effective.

1. The motivation should be brief.
2. The motivation should not be overemphasized. It should be a means to an end, not an end in itself.
3. The motivation should elicit the aim of the lesson from the class. This is a fine way of determining how effective the motivation actually is.
4. The motivation should be appropriate for the class's level of ability and interest.
5. The motivation should draw on motives actually present in the learner.

Remember, getting students interested in the lesson makes an enormous difference in how much they learn. Planning your motivational devices carefully is a critical part of every lesson!

Chapter 3

The Art of Questioning

Ever since the age of Socrates, Aristotle, Plato, and other famous teachers, the art of asking good, clearly focused questions has dominated the role of the teacher. Asking questions may come naturally to some, but the type of questions and what the teacher follows the answers with could determine the success or failure of a lesson. Knowing what questions to ask is definitely an art and, like any art form, can be developed with practice. With conscientious preparation and patient practice, proficient questioning skills can be cultivated which not only enliven and enhance engagement in the classroom, but even have a substantial impact on the entire teaching-learning process. The great impact of well-versed question asking on students' education is why teachers must consciously develop habits of questioning that will strengthen their teaching performance.

In this chapter, we present both math coaches and math teachers with the many aspects of effective classroom questioning, including the delivery of questions,

- defining motivation behind questions asked,
- elements of productive rather than ritualistic questioning,
- tailoring questions to group size and ability,
- avoiding questions which mention specific biases against the students' backgrounds,
- the pitfalls of bad questioning with a guide on how to counter them.

3.1. Direct and Simple Language

Classroom questions should be direct and simple in language to engage student focus on a question's content, not on the language used in its delivery. That is, if the language distracts students from the content, perhaps by being either too complex or too humorous, the potential effectiveness of a question may be lost. By using direct and simple language (i.e., appropriate for the level of the intended class), a teacher can employ classroom questions to fulfill their desired function. Also, be certain that the mathematics of the question is correct. Sometimes in order to make the language easily understood, the mathematical accuracy is compromised.

3.2. Definite and Clear Meaning

Classroom questions should also be definite and clear in meaning. If a question lends itself to various interpretations, students may be reluctant to respond. So, to maximize the number of volunteer respondents, such ambiguities should be avoided. Shorter questions are often less confusing. For example, asking the class to decide what happens if we replace the variable, x, by 4 in the expression $\frac{5x+6}{x-4}$ is a good question, and should lead to a discussion of division by zero. This question is definite and clear as to what is expected — it calls for only one or two points in a line of reasoning. A teacher should ask more questions rather than try to limit the number of questions and lengthen each one. By trying to ask too much in a single question, the teacher may become prone to asking multiple or overlaid questions.

3.3. Logical Sequence

Failure to consider expressing questions in a logical sequence can lead to disinterest, as students struggle to follow a lesson's structure. An inexperienced teacher's impatience with the developmental process may cause him or her to rush to the pivotal (or prime) question of a lesson, without spending enough time leading up to it with shorter preparatory questions. This impatience often diminishes the ultimate effectiveness of the pivotal question, which otherwise would have elicited the highlight of the lesson. Thus, teachers should give special attention to all parts of a line of questioning that develops a train of thought in logical sequence. This means that the same care should be given to the early, perhaps almost trivial (or review) questions as is given to the pivotal and culminating questions. Remember, the pivotal question is rarely effective if it is not properly built up through a carefully developed sequence of subordinate questions in predetermined order.

3.4. Questions Keyed to Class Ability

The level of ability of a class should determine the language and complexity of the classroom questions used. It is easy for a teacher to use the same questions during two consecutive class periods, especially when the same lesson is involved. Yet if the ability levels of the two classes are different, this practice should be consciously avoided. For the slower or perhaps less sophisticated class, a simpler, more broken-down delivery might be used than with a class comprising more able students. Teachers should be careful not to be condescending, but at the same time they should not conduct the class with language beyond the students' ability to grasp the content comfortably. By asking questions properly suited for the intended audience, teachers will improve communication with classes at all levels.

3.5. Questions That Stimulate Effort

Questions should arouse an effort. While gearing the phrasing of such questions to the appropriate level for a class, teachers must make a special effort to prepare questions that are sufficiently challenging to intrigue yet not too difficult to stifle the class. Good classroom questioning should hence, create a stimulating atmosphere throughout the lesson. These questions should be short and crisp, and arranged in a logical sequence that builds to the desired point. A sequence of questions might well consist of a blend of factual and thought questions, with a majority of the latter kind. Included would be a balanced mixture of some short yet challenging questions along with other review or connecting questions. Such a blend should go far to stimulate active learning.

3.6. Open-Ended Questions

Open-ended questions allow students to reach conclusions and make mathematical decisions consistent with their understanding and development. On a class examination, students can demonstrate the kind of depth that is impossible to determine on the basis of choosing one of several multiple-choice items or writing a single-number response. Pupils may also arrive at multiple possible "correct" answers to open-ended questions, encouraging thinking and reasoning. For example, in a workshop setting, ask your teachers or students to determine which is larger, $-x$ or $+x$. This is far more desirable a question than asking if -3 or $+3$ is larger. This method of questioning naturally leads to healthy and unrestrained discussion that neither "True or False" nor multiple-choice type questions will achieve.

3.7. Maintaining Student Interest and Attention

Coaches should encourage teachers to keep in mind that classroom questioning is an effective way to hold student interest throughout a lesson. A few important points to be emphasized regularly are as follows:

1. Every effort should be made to call on as many different students throughout a lesson as possible and to try to avoid predictability in calling on students.
2. Mixing calling on non-volunteers with calling on volunteers will keep everyone attentive.
3. Beginning a course by requiring students to give complete answers to classroom questions will help to ensure that such a practice will eventually become habitual in the class.

It is a good habit to praise students (with tact) for correctly answering a question. It is equally important, however, for a teacher to handle incorrect answers constructively and not insult students.

The best treatment for an incorrect answer depends on the type of class and the time available in the lesson. Under no circumstances should the teacher be abrupt or scold a student for giving an incorrect answer. This would likely have an adverse effect on the student's learning and inhibit future willingness to respond to the teacher's questions.

With classroom time permitting, the teacher may guide the student to realize an error through a series of questions specially tailored to this discovery. Alternatively, the teacher may choose to refer the question to the rest of the class.

It is not unadvisable that students attempt to answer questions posed by other students, and teachers should not feel that only they should provide the answers. Peer interaction can produce interesting results. For example, by answering another student's question, the first student can better learn the concepts discussed. The process of teaching usually allows the teacher to understand better the subtleties of subject matter being taught. The same can be true when one student explains a concept to a classmate – both get a better understanding of the concepts under consideration.

When students are expected to respond to one another's questions, there is a greater degree of alertness throughout the classroom, because students cannot predict precisely when they will be called on to correct or answer another student. This alertness could go far to enliven the class.

3.8. Avoiding Repetition of Questions

The teacher's question should generally not be repeated. Naturally, if for some unusual reason the question was inaudible, then repetition may be necessary. Having a student repeat a question when it was not heard by some students can even introduce variety. Nonetheless, habitually repeating a question can make the class inattentive, because they can rely on the question being repeated and continually call for repetition to waste class time. But if they know that the repetition of a question is not easily obtained, this scheme will fail, resulting in an attentive class with no time wasted.

Sometimes a teacher may experience self-doubt, and having asked a question of the class, may repeat the question in perhaps a rephrased form before the class has even had a chance to respond to the question. An uncertainty on the part of the teacher about the clarity of the question will provoke this immediate repetition of the question. Very often only the teacher finds the question unclear. The class may have been ready to respond to the original question, but after hearing the rephrased question may be somewhat confused. The teacher should let the original question stand, give students sufficient time to respond, and rephrase the question only if no correct response is forthcoming. Students often have an unsuspected ability to interpret a teacher's question properly, even if it is a bit unclear. At such times teachers should not be overly critical of their own questions. To prevent this entire situation, proper questioning preparation should be practiced.

3.9. Avoiding Repetition of Student Answers

The teacher should not repeat student responses, for reasons similar to those discussed earlier. If students can rely on the teacher to repeat the important student responses to teacher questions, they may eventually not even listen to their classmates, which will greatly inhibit active student interaction throughout a lesson. If, in the teacher's judgment, a student's response is inaudible, then the teacher should have that student or another student repeat the response. Consistency with this procedure will eventually cause students automatically to speak loudly and clearly just to avoid having to repeat their responses (or hear other students do so).

Some teachers have a habit of "mentally processing" a student's response aloud. This results in a repetition of the student's response. Most teachers, when made aware of this habit, can curtail it. Thus, recording a teacher's lesson and reviewing it will be helpful in demonstrating this flaw in a teacher's performance. A teacher who cannot avoid such repetitions, however, should at least try to incorporate this repetition into the next statement or question. This way it may not appear as a simple redundant repetition.

Teacher repetitions of student responses are also caused by the teacher's fear that unless a substantive statement is made by the teacher, the class will not note it properly. This, too, will only be so if the teacher allows it to happen. The teacher sets the class tone and routines and the manner in which classroom questioning and the resulting student responses are handled.

3.10. Calling on Students

Another way to generate constant student interest is to call on a particular student for a response after a pause at the end of a question. If the class is accustomed to having the teacher call on a student to respond to a question after it has been asked, each student will be attentive just in case he or she is the one selected by the teacher to respond. On the other hand, if a teacher addresses David before asking the question, perhaps only David will be attentive, because the rest of the class will know they are not being asked the question.

This latter situation does not promote an active involvement in the learning process. Thus, the teacher should address particular students at the end of the question and thereby, ensure student attention throughout the lesson.

It is sometimes helpful to discover which students in the class are trying to avoid being called on. There are times when it may be wise to call on them or simply meet with these students after class to discuss their apparent avoidance. The question is, how can the teacher discover which students are trying to avoid being called upon? Experience has shown that if after the teacher asks a question of the class and pauses to look around for a student to call on for a response, a student appears to make a conscious effort to avoid eye contact with the teacher, then the teacher can assume that student does not want to be called upon. This avoidance sometimes manifests itself in the student's appearing to be very "busy" so that the teacher's calling on him would be seen as interfering with his "concentration." Sometimes the teacher can discourage this by making a general statement to the class about this type of avoidance. Done properly, the class may find this humorous (because many of the students may have been guilty of this type of behavior at one time or another) and realize that this teacher must be very clever and not easily fooled. However, the teacher then must be watchful for other, perhaps more creative, techniques that some students may use to avoid questions (especially when they feel they cannot answer them properly).

3.11. Wait-Time after Asking a Question

Allowing students sufficient time to think about a teacher-posed question is a very important aspect of classroom questioning. One of the leading researchers in the questioning behavior of teachers is Mary Budd Rowe. Her findings over the years have had a significant impact on teacher performance in the classroom. In her extensive analysis of classroom performances, she has found that most teachers, on average, wait less than one second for students to respond to their questions. On the other hand, some teachers wait an average of three seconds for students to reply. When she compared student responses with different wait-times, she found that the longer wait-times (three seconds or more) produced more thoughtful responses, increased classroom discussion, and enabled students to analyze a situation more critically than did the shorter wait-times following the teacher's questions. One effective technique to determine the wait-time following class questions is to make a recording of your teacher's lesson and then, during the playback, to time the periods of pause following each question. Try helping your teachers with an increase of their wait-time, if it is too short. Again, record the lesson to inspect for an increase of wait-time. Such an exercise ought to produce favorable results.

Once you have succeeded in increasing the wait-time following your teacher's classroom questions, you might have them try to pause briefly after a student has responded, to allow a moment for reflection or to permit a student to add more information to his first response. This second type of wait-time has similar effects on the learning environment as the first type of wait-time. This has been shown by an analysis conducted by Dr. Rowe of more than 800 recordings of lessons in urban, suburban, and rural schools. (For further information on Dr. Rowes's research findings see: Mary Budd Rowe, "Wait Time: Slowing Down May Be A Way of Speeding Up!" Journal of Teacher Education, Jan.–Feb. 1986, pp. 43–50.)

3.12. Variety in Questioning

Perhaps one of the most important elements in good classroom questioning, as in most aspects of good teaching, is variety. Variety can refer to the types of questions asked, to the manner in which questions are asked, to the way students (volunteers and non-volunteers) are called upon to respond to questions, and to the procedure by which responses are handled. Variety reduces predictability, which in turn ought to promote continuous stimulation. When teachers vary the types of questions asked, students are required to be alert to more than simply the content of the questions. This additional alertness should have a refreshing spin-off toward improved learning. In addition to providing students with a more interesting experience, teachers also are more apt to be energized by the challenge of creating continuous variety in their questioning.

3.13. Ten Types of Questions to Avoid

In any sequence of questioning, a few weak questions may be asked without harm, but many poor questions will undermine a lesson's effectiveness. Following are ten types of questions a teacher should consciously avoid, for they may be counterproductive.

3.13.1. *Overlaid questions*

In the midst of asking their classes a question, teachers often find that the question is not specific enough to elicit the desired response. Rather than let the original question ride on its merits and give students a chance to answer it, teachers may augment the question before students have even had a chance to answer the original question (recall section 3.8). When this happens, students who may have understood the original question may now hesitate to answer, because they are uncertain about their understanding of the entire question. Thus, by elaborating on a question they felt was unclear, teachers may have caused confusion by tagging on an additional thought. One such overlaid question could be: "What method shall we use to solve this problem, and that will make our solution elegant?"

Even if a student knows which method to use to solve the problem in question, he may avoid answering the question because of uncertainty about the second part of the question, namely, whether his method will produce an "elegant solution." An improved way to ask this question is: "What method shall we use to solve this problem (pause), Barbara?" followed by "What would be a good first step in this solution, (pause) Max?" In this example of an overlaid question, an elaboration of an original question was tagged on. This had exactly the opposite effect of what was intended.

3.13.2. *Multiple questions*

Multiple questions are formed by asking two related questions in sequence without allowing for a student response until both parts of the question have been asked. Consider the following: "Which triangles should we prove congruent, and how will that help us prove $\overline{AB}$ parallel to $\overline{CD}$?" Although a student may know which triangles need to be proved congruent, he may not know how congruent triangles will help prove $\overline{AB} \parallel \overline{CD}$. Because of this, the student will probably not answer. However, if the question is asked in two parts, allowing for an answer to the first part before the second part is asked, then more students will be likely to respond. This may be done as follows: "Which triangles should we prove congruent (pause), Jack?" Then the teacher may ask "How will these congruent triangles enable us to prove $\overline{AB} \parallel \overline{CD}$ (pause), Sam?"

This type of question is similar to the overlaid question in that it also has two parts. It differs from the overlaid question by being two questions that could actually stand separately. Teachers often resort to multiple questions when they feel that the time remaining for a lesson is too short, or when they get somewhat impatient and want the lesson to move along more rapidly. As before, students can easily be discouraged from answering this type of question. To provide a correct response, a student must be able to answer correctly *both* parts of the question. Thus, a student who can correctly answer only one part of the question will not volunteer to respond. By reducing the pool of students who will respond to questions, the teacher diminishes active learning throughout the class. Consider the question: "What is the discriminant of this equation (pointing to a quadratic equation), and what type of roots does it have?" This multiple question could stand as two separate questions, but in its present format it will most likely discourage students from responding. The same information can be elicited more effectively by asking: "What is the discriminant of this equation (pointing to a quadratic equation) (pause), Lisa?" followed by "Based on the value of the discriminant, what type of roots does this equation have (pause), Charles?" By stifling student responses, multiple questions reduce the effectiveness of a lesson and should, therefore, be avoided.

3.13.3. *Factual questions*

There is certainly nothing wrong with asking a question that has a simple factual response if the question is part of a buildup of a series of sequential facts necessary for the solution to the problem under consideration. Otherwise, however, isolated factual questions do little to stimulate student thinking. Consider the question: "What is the Pythagorean theorem?" Not much thought is required to respond to this question. A student either knows the answer or doesn't. If we agree with our original premise regarding classroom questioning, then aside from being a part of a sequence of questions, factual questions contribute little to an active learning environment.

3.13.4. *Elliptical questions*

Questions that are unclear because the teacher has omitted specifics offer nothing to a lesson. Although not particularly harmful to a lesson, an elliptical question is simply an unnecessary waste of time.

One such is the question: "How about these two angles?" Teachers frequently have a habit of thinking aloud. They may be looking at a pair of angles and thinking of what to ask about them, such as "What is their relationship?" or "Which angle has the greater measure?" In either case, the teacher may instead first voice the thought: "How about these two angles?" Verbalizing this thought into a question that has no answer wastes classroom time. The teacher might have asked, "What is the relationship between these two angles (pause), David?" Here, a definite answer was required. Had the teacher wanted to say something (so as to avoid a lull in the lesson) when he was thinking about the two angles, he could have said, "Consider these two angles." This would have served the intended purpose and not wasted classroom time with possible student wisecrack answers such as "How about them!" Similarly, the question "What about these two parallel lines?" has no real answer.

As before, this elliptical question asks for either nothing or more than most students are prepared to offer. At any rate, because of the omission of specifics, it leaves itself open to wisecrack responses. The teacher may wish to say something like "Which angles can we prove congruent using these parallel lines (pause), Laura?" The teacher does not have to overreact to avoid lulls. Instead, she should stop and give some thought to a question, rather than ask it in a form that has no clear response.

3.13.5. *Yes-No or guessing questions*

For the most part Yes-No or guessing questions have little value. With few exceptions a Yes-No question — such as the following — can easily be transformed into a good thought question. A student attempting to answer this question takes a very small risk as his chance of merely guessing the correct answer is actually 50 percent. The teacher asking this sort of question more often than not is seeking affirmation. In addition, this diminishes the entire premise of the question, and makes it redundant and rhetorical. This question might be improved to read, "What is the relationship between $\overline{AB}$ and $\overline{CD}$ (pause), Jack?" This would require the student to explore the possible relationships that two line segments may have and then answer appropriately. In its furnished state, the question provokes active learning among the students. Another such example is: "Is triangle *ABC* isosceles?" Why would a teacher ask this question if the triangle were, in fact, not isosceles? Unless the teacher were set to trick the class, students would be correct in assuming that the teacher is simply seeking an affirmative response. Why then ask the question? The question will be far more productive when asked as "What type of triangle is *ABC* (pause), José?" It is therefore a good practice to avoid Yes-No or guessing questions whenever possible. Sometimes a Yes-No question has some value, if it is followed up by a further probe such as "How do you know?" or "Why is that so?"

3.13.6. *Ambiguous questions*

Occasionally, a teacher may seek a response that requires a specific interpretation of a situation. Here the questioner, trying to get the desired response with one question, may likely ask an ambiguous question, one that can have a variety of different, yet correct, answers. The desired response would be more easily attained by asking a series of short sequential questions.

"Here we can use the following as an example: "How does the law of sines differ from the law of cosines?" Many different correct answers can be given to such a question. Certainly, the context in which this question is asked will help narrow the choices among the correct responses. Students will tend to shy away from responding to this question, however, out of obvious confusion caused by its ambiguity. Students may wonder if the question refers to the difference in appearance of the two laws, the difference in application, the difference in derivation, and so on. One possible form in which this question may be improved is "Under what different circumstances are the laws of sines and cosines used (pause), Henry?" Because such confusion is obviously counterproductive, ambiguous questions should be consciously avoided in classroom questioning.

Consider another question: "What is the relationship between the area of a circle and the circumference of a circle?" Once again, this question has many correct answers. Is the questioner concerned about the numerical relationship, the physical relationship, the dimensional relationship, or some other, less obvious relationship? Although the context in which the question is asked will assist students in responding to the question, confusion is rarely avoided when an ambiguous question is asked. One specific way of asking this question is "What is the ratio between the area and circumference of a circle (pause), Gabriella?"

However, it is not necessarily a bad or undesirable feature to pose questions having several correct answers, perhaps with the intent of generating discussion. We are considering the "ambiguous question" here rather than this type of question. Before asking a question that is likely to be ambiguous, have the teacher set the specifics about the situation, then ask short, simple questions to elicit the desired response.

3.13.7. *Chorus response questions*

Although a question calling for a chorus response may be good, the chorus response often provides little value to the lesson. When a class responds in chorus to a question, the teacher usually cannot determine which students are answering incorrectly and which students are not responding at all. Furthermore, a chorus response can become too unclear for students eager to learn from the answer to hear the answer correctly. By missing the answer to a question, a student could be missing an important link in a chain of reasoning, resulting as a detriment to the learning process for this student.

Consider: "What type of quadrilateral is *ABCD*, class?" If we assume that not everyone in the class knows the correct answer to the question, some students will be shouting an incorrect answer while others will be answering correctly. Should one student not respond but instead listen for the correct answer, he may hear an incorrect answer (because a wrong response may have come from someone close by) and then try to learn a concept with an incorrect piece of information. The time lost in correcting this error is certainly undesirable. Furthermore, a preponderance of chorus response questions will allow some students to glide through the lesson without actually learning the subject matter presented. The teacher in this situation will be unable to detect individual difficulties, because they are likely to be clouded by the chorus responses — another reason to avoid the chorus response question as much as possible.

But an occasional use of this type of question may be acceptable if the response is not too crucial to a development and if it is necessary to involve the entire class, even for the sake of variety. A change of style, offering variety to the lesson, may be a healthy feature. Even when serving this purpose, however, the chorus response question should be used sparingly.

3.13.8. *Whiplash questions*

A whiplash question is one that is usually not planned by the teacher. It comes about when a teacher decides to make a question out of a statement midway through, such as: "The slope of this line is, what?" Aside from possibly frustrating students, little harm is caused by this type of question. Perhaps its greatest flaw is its uselessness. Not expecting a question, the students are caught off guard. They must first mentally rephrase the question before attempting an answer. Under normal circumstances the beginning key word of a question (e.g., why, when, what, how) puts the students in a psychological set, ready to receive and process a question. The whiplash question does not provide this readying process and thereby wastes time and loses much of the student audience.

More productive ways to ask such a question would be: "What is the slope of this line (pause), Richard?" On the other hand: "We now have $\overline{AB}$ parallel to $\overline{CD}$ because of which theorem?" This question would have been much more effective had the key word announcing a question been at the beginning of the question. The original question would then read, "Which theorem justifies the fact that $\overline{AB}$ is parallel to $\overline{CD}$ (pause), Ceri?" In this form, the students know from the first word that a question is being asked, while the second word has them focus on the various theorems learned while they listen to the remainder of the question. At the completion of the question they are ready to respond without wasting time to rephrase the question.

While the latter form of question-asking is clearly more efficient than the whiplash format, a teacher need not necessarily turn every statement into a question just for the sake of producing student participation. Such forced attempts could easily become counterproductive.

3.13.9. *Leading questions*

A leading question is one that tugs the desired response from the student and serves no reasonable function. Consider the question "Wouldn't you say that ΔABC is equilateral?" Most students would be quite reluctant not to agree with the teacher. Thus, the question does not provoke much thought because the student is more than likely to simply agree. Or consider the question: "Seven is a factor of 35, isn't it?" Again, there is no need to turn the statement "Seven is a factor of 35" into a question. The teacher would be better served either by leaving the statement stand as is or by asking a question such as "What are the factors of 35 (pause), Evelyn?" or "By what number must 7 be multiplied to yield 35 (pause), Barbara?" Each of these questions requires some thought on the part of the students before they answer. In addition to replacing a time-wasting question, each induces active learning.

3.13.10. *Teacher-centered questions*

It is generally desirable to have students consider the teacher as part of the class. Although students are well aware of the difference in roles, it is more effective for the teacher to use the first-person plural (i.e., *we* and *us*) when addressing the class, if appropriate. For example, saying "Let us consider the following..." rather than "I have the following..." would make the class feel that they are all part of one group working together on a common problem. They do not need a constant reminder that they are the students and the teacher is distinct from them. Regular use of the first-person singular (i.e., *I* and *me*) could create an invisible barrier between the teacher and the class, a possible detriment to a healthy, active learning environment.

For example, a more enticing way to ask "Give me the solution set of $3x - 5 = 2$" is "Give us the solution set of $3x - 5 = 2$ (pause), Dan."

Similarly, rather than ask: "What must I do next to solve this problem?", inquire instead," it would be better to ask: "What must we do next in solving this problem (pause), Lauren?"

Each of the preceding examples illustrates the sort of teacher comment that seeks to set apart (albeit subconsciously) the teacher from the students, which is not particularly conducive to a good classroom learning environment. You should be certain to emphasize that their role as a teacher includes being an active member of the class.

3.14. Classroom Questioning as a Means to Generate Higher-Order Thinking

Questioning in an optimally interactive classroom is used by the teacher to help students gain understanding and by students to obtain guidance that will assist them to clarify ambiguities and resolve confusion. Traditional guidelines for questioning deal exclusively with the teacher's questions, setting down formal guidelines for what is acceptable and what is not. In the following paragraphs, several such issues will be raised based on student psychology as it is now understood.

For example, not all ambiguity is bad. We will explore ways in which ambiguity might be employed consciously, on occasion, to assist students in acquiring a deeper understanding of familiar material.

Some updated formats for traditional questions are likely to reduce student anxiety without in any way lessening the quality of learning, while other new techniques are powerful in their own right. Accepting new ideas authenticates older practices — it builds on the body of new knowledge about students, the way the brain works, and the routes by which skills and understanding are incorporated into long-term memory. It recognizes pedagogy as a living art and science.

3.14.1. *Explain your thinking*

There is much value in questions that probe the rationale for a student's answer. Every student should have a reason for each response given; the accessibility of rationale assists the student in reconstructing correct answers and storing new information in long-term memory. Unfortunately, many people, both young and old, find "why" questions threatening. A student has correctly answered the challenge to compute -5×-4 with the answer $+20$. The teacher then asks, “Why is positive 20 the correct answer?” The student panics and responds, “The rule says that when you multiply two signed numbers with the same sign, the answer must be positive.” Is that what the teacher wants to hear? If so, the teacher is encouraging rote learning. If not, exactly what does the teacher now say to correct the student’s correct but undesired response?

A superior psychological and pedagogical technique would be to follow up with the challenge to “Explain your thinking.” To the student response noted in the previous paragraph the teacher might say, “That is not your own thinking. Explain why *you* think the product of two negative numbers must be positive.” A second advantage to asking students to explain their thinking is that most people find it far less threatening to talk about what *they* think than to try to explain what *someone else* thinks. The last advantage to “Explain your thinking” is that it pinpoints the best route to long-term memory, namely, the reconstruction of knowledge resulting from the student’s own thought processes.

3.14.2. *Compare and contrast*

An earlier section (13.6) cautions against ambiguous questions like “What is the relationship between the arc of a circle and the circumference of a circle?” The student may become confused because this question, as previously noted, has many answers. The value of the question can be preserved and heightened by using the unambiguous *compare and contrast* format.

For example, “Compare and contrast the circumference and the area of a circle.” All of the student’s insights are called for. The students might compare circumference to area by citing that both measures are associated with the circle, involve the number π, depend on the radius of the circle, etc. The student might contrast circumference and area by noting that circumference is a measure of (arc) length, whereas area is a measure of the extent of a region, or by noting that the area can be deduced from the circumference by extending the formula for the area of a regular polygon, $A = \frac{1}{2}ap$ (where a = apothem and p = perimeter) or by noting that for a circle, $A = \frac{1}{2} \times r \times 2\pi r = \pi r^2$.

"Compare and contrast the ways you think about adding fractions and multiplying fractions." Students would be able to both cite the algorithms *and* discuss the rationale behind the algorithms. Students might be led to consider the problem of adding and subtracting any expressions involving units (because, after all, a denominator is a unit). A student who could insightfully compare and contrast the rationale or impossibility behind adding 5 pounds to 4 feet, multiplying 5 pounds by 4 feet, adding 5 inches to 4 feet, and multiplying 5 inches by 4 feet is well on the way to understanding both mathematics and science more fully.

By asking *compare and contrast* questions (also a form of open-ended question) such as these and avoiding specific questions that tend to incite ambiguity, the teacher creates excellent opportunities for open discussion and to receive answers to questions he or she wouldn't have thought of asking.

3.14.3. *Prove or disprove*

Students will benefit from being exposed to the *prove or disprove* format early in their introduction to mathematical proofs. The direction to *prove* tells them that the statement they are given holds in general. To be truly adept at this, students must realize that most statements they formulate themselves will not be provable. The *prove or disprove* format encourages students to try to find a counter-example to an unknown assertion *before* trying to find a proof.

For example, prove or disprove that any quadrilateral with diagonals that are both perpendicular and equal in length is a square.

A simple diagram of a kite would serve as an adequate disproof. A kite can have equal diagonals which are perpendicular, but they do not bisect each other, whereas were they to bisect each other the resulting quadrilateral would be a square. In this case, the disproof by example is rather clear.

However, students must bear in mind that even a pattern that appears to be consistent is not sufficient to prove something true for all cases as you can see from the following example:

$$3 = 2^0 + 2$$

$$5 = 2^1 + 3$$

$$7 = 2^2 + 3$$

$$9 = 2^2 + 5$$

$$11 = 2^3 + 3$$

$$13 = 2^3 + 5$$

$$15 = 2^3 + 7$$

$$17 = 2^2 + 13$$

$$19 = 2^4 + 3$$

$$\vdots$$

$$51 = 2^5 + 19$$

$$\vdots$$

$$125 = 2^6 + 61$$

$$127 = ?$$

$$129 = 2^5 + 97$$

$$131 = 2^7 + 3$$

Here you will note that it appears from the first few examples that odd numbers can be expressed as powers of 2 plus a prime number. Until the number 125, all is well. However, with the number 127 it is no longer possible. This example shows that a pattern alone does not constitute a proof. These are the sorts of examples that a math coach should be exposing teachers to so that they can share this thinking with their students.

3.15. Developing Engaging and Valuable Questions

With the change in focus of the mathematics curriculum and an emphasis on problem-solving and reasoning, the kinds of questions asked become even more important. After all, asking appropriate questions can determine whether or not the material being taught is understood. The response to asking a question such as "The answer is an isosceles triangle, isn't it?" isn't very valuable. Who would want to contradict the teacher? Many teachers inappropriately summarize a lesson by asking "What have we learned today?" This is another question to be avoided; it just isn't specific enough to elicit a meaningful response. To help your teachers refine their arsenal of questioning techniques, below are procedures you may employ for developing questions.

Developing a plan is key — when meeting with one of your teachers, help that teacher design a sequence of questions to attempt in each lesson. For example, a first question might be "What operation would you use to resolve the problem?" After calling on a student to answer, follow up with a simple "Why did you choose that operation?" Finally, "How can we check to see if that operation chosen is correct?" Notice that the actual answer the student gave was not evaluated as correct or incorrect. Rather, the original question leads to an insight into the rationale for the student's choice. The key question is, of course, "Why?"

An excellent planning technique to demonstrate how questioning can be key to problem-solving is to arrange a workshop/meeting with several of your teachers. Treat it as a brainstorming session – present the group with an interesting problem and ask them to work together in small groups to develop a series of questions they would use to solve the problem.

The following is an interesting sequence of carefully and creatively planned questions: The school principal came to my house to enroll my children in school. I told her I had three children. When she asked me for their ages, I told her, "All I could say was that the product of their ages is 72 and the sum of their ages is the same as my house number." She looked at my house number and said: "I still can't tell!" I told her I had forgotten to tell her that the oldest likes blueberry muffins. With that, she wrote down their ages and left. How old are my children? (Consider only integers.)

1. How many children do I have? (3)
2. What do we know about their ages? (The product is 72)
3. What else do we know about their ages? (The sum of their ages is the same as my house number)
4. What might we do first? (List all triples whose product is 72.)

1, 1, 72	2, 2, 18	3, 3, 8
1, 2, 36	2, 3, 12	3, 4, 6
1, 3, 24	2, 4, 9	
1, 4, 18	2, 6, 6	
1, 6, 12		
1, 8, 9		

5. Why is there no 1, 5, X nor 1, 7, Y? (because 5 and 7 are not factors of 72)
6. What should we do next? (Add the ages in each triple — same as house)

1 + 1 + 72 = **74**	2 + 2 + 18 = **22**	3 + 3 + 8 = **14**
1 + 2 + 36 = **39**	2 + 3 + 12 = **17**	3 + 4 + 6 = **13**
1 + 3 + 24 = **28**	2 + 4 + 9 = **15**	
1 + 4 + 18 = **23**	2 ++ 6 + 6 = **14**	
1 + 6 + 12 = **19**		
1 + 8 + 9 = **18**		

Notice that all the material has been developed through specific questions.

7. Why didn't the principal know their ages? She knew the house number. (Because there are two house numbers of 14.)
8. How did we "break the tie"? (The *oldest* likes blueberry muffins.)
9. What are the ages? Only (3, 3, 8) has an oldest, and a sum of 14, which the other option (2, 6, 6) which has a sum of 14 has no unique oldest. The sought ages are, therefore, (3, 3, 8) years old.

Notice that there are additional questions that can be asked. The concept of 5 and 7 not being factors of 72, and the idea that a factor is a divisor of the number are the crucial mathematical concepts being developed, all through a careful set of questions.

3.16. Some Final Do's and Don'ts of Student Questioning

Some questioning techniques are quite easy to master with constant practice and suggesting these to your teachers can usually lead to an understanding of the questioning techniques.

1. Don't specify the name of the student to be called on *before* asking the complete question. If the teacher says "John, why did we multiply instead of divide?" every other student in the class can relax as they're not involved from the onset. A better technique to emphasize with your teachers is to simply ask the question, take a brief pause (several seconds) and then name the student from whom you want an answer.
2. As said before (in 13.7), don't direct a question to the whole class and look for a chorus answer. When many voices join in the response, the question and its answer lose significant effectiveness. Some students will not even bother to answer; others will join whichever group shouts the loudest. In either case, the teacher will not get any evaluation of student understanding from the question.

3. Do try to follow up a student's answer by asking "Why?" This eliminates guesswork and helps the weaker students determine how the answer was arrived at. We cannot emphasize enough the value of the simple question "Why?" Teachers should not repeat a student's answer for the benefit of the class. The students will quickly learn that they need not listen to each other, but should wait for the teacher to talk. If an answer needs repeating, teachers should call on another student to repeat the answer. This will serve well to insure uniform attention to all aspects of the lesson.
4. You should impress upon your teachers not to use questions for punitive measures. Calling on a student whom they know will not give the correct answer is rarely valuable. They are liable to provoke an encounter of sorts, if the student gets frustrated. Embarrassing a student will get the teacher nowhere, and could simply disrupt the lesson.
5. Teachers should avoid elliptical or vague questions. They shouldn't ask "What about an isosceles triangle?" or "Tell us about prime numbers." The should be more specific! Teachers should ask the question in a form so that the students know what the teacher seeks to determine. For example, a student could be asked to define a prime number, or describe what properties of a number determine a prime number?
6. Try to find something positive about a student's answer, even if it's incorrect. Say something such as "That's a good start. Well, you are partially right." or "Who can build on Maria's answer?"
7. One should not preface a question with an evaluation, such as "Here's an easy one: Name a consecutive prime triple." If the student recognizes that the answer is 3, 5, 7, then yes, it's easy. But think about how the student who doesn't know the answer must feel if that is meant to be *an easy one.*
8. One should not ask a question which contains the answer. For example, it is not productive to ask, "This is an isosceles triangle, isn't it?" It's better to simply state "This is an isosceles triangle."

To summarize the effective kinds of questions, let's examine the following list.

(a) Effective questions are asked at an appropriate level for the students. Be certain that they understand the question.
(b) Effective questions are aimed at one specific purpose or idea. Don't ask for generalities or about generalities.
(c) Effective questions are brief. The lengthier the question, the more likely the students will have difficulty to answer it correctly.
(d) Effective questions are designed to be thought provoking. The simplest question ("Why?") can elicit a sustained response and generate thinking.
(f) Effective questions are asked one at a time. Don't follow up a question with a second question before the first one has been answered. For instance, don't ask "Is this a parallelogram? What kind of parallelogram is it?" The second question essentially answers the first question — perhaps before the student has had a chance to respond.
(g) Effective questions are not repeated by the teacher. Since if you are going to repeat the question, why should students listen the first question?
(h) In coaching teachers, have them keep in mind the importance of questioning. Suggest to teachers that they prepare a few key questions for a lesson and have them discuss these questions with you, their coach, beforehand. Help them to see how effective questions can go a long way in the development of a productive learning environment. Remember, like any skill, questioning is best learned by careful repetition and practice.

Chapter 4

The Role of Problem-Solving

Over the years, problem-solving has emerged as one of the major concerns at all levels of school mathematics. As far back as the year 2000, the National Council of Teachers of Mathematics (NCTM) stated rather boldly in their *Principles and Standards for School Mathematics* that "**Problem-solving is not only a goal of learning mathematics but also a major means for doing so.**" The Council goes on to say that problem-solving "**strategies are learned over time, applied in particular contexts, and become more refined, elaborate, and flexible as they are used in increasingly complex problem situations.**" We are in complete agreement! In fact, we would go one step further: We feel that problem-solving is not only a skill to be taught and used in mathematics, but also one that will be carried over to everyday "problems" or decision-making situations, thus serving a person well throughout life. Even though most math teachers are thoroughly convinced that they focus on problem-solving, they typically only consider the textbook exercises as problems, whereas this concept of problem-solving far exceeds the repetitive exercises in a textbook. It is thus, sad to say, that many math teachers are simply not as aware of as many problem-solving strategies as they ought to be. A major responsibility of the math coach is to incorporate genuine problem-solving strategies into their regular teaching.

In many cases, students seem to feel (perhaps because of earlier experiences) that a problem can only be solved in a single way, specific to the "type" of problem being taught. Many state curriculum guides are now encouraging students to "think out of the box," and we offer ideas that the math coach can use to encourage the use of more creative problem-solving strategies. Teachers, however, may not be aware of the numerous problem-solving strategies that can be used to provide efficient and elegant solutions, hence unintentionally conveying to their students the notion that problems can only be solved using a particular approach. The math coach should therefore, be a resource for alternative methods.

In this chapter we will examine **ten strategies**, which are widely used in problem-solving both in mathematics and in real-life decision-making situations. In the classroom, these strategies provide an alternate plan for resolving many problems that arise within the curriculum. We will illustrate these strategies with appropriately selected problems that teachers will enjoy and adopt. To best facilitate this, we recommend a careful review and study of the examples provided so that the strategy becomes a genuine part of the teacher's thinking process or, one might say, a part of their arsenal of problem-solving tools.

To provide the math coach with a true command of problem-solving strategies, we will be presenting illustrations that span the entire school curriculum. While it is true that many problems can be solved using the tried-and-true techniques of algebra and geometry (and we do show these solutions as well), the purely "mechanical" approach often masks some of the efficiency, beauty, and elegance of the mathematics. In many cases, the problem-solving strategies presented here as alternative methods can make the solution of a problem much easier, much neater, and much more understandable and, thereby, enjoyable!

As these strategies are already widely used subconsciously in normal decision-making, we aim to show how they are formulated and purposefully applied to mathematics as well. This carryover into life outside school adds value to the mathematics our students study, and will ultimately benefit them holistically. The math coach enhances this effect by appropriately customizing the ideas and concepts presented here for their audience to best utilize.

To do this, the math coach should infuse enthusiastic feelings and attitudes into the teachers' instructional routine, motivated by the understanding that problem-solving is the cornerstone of any successful mathematics program. The ultimate goal is to improve not only the students' outlook toward mathematics, but also their skills and abilities.

One should, however, not lose sight of the importance of the topics being taught while imparting soft skills. Building a solid foundation in the understanding of mathematics — clearly the task of all mathematics teachers — is still the primary objective. We see the thematic approach to teaching mathematics nicely complemented by the emphasis on problem-solving throughout the instructional program, and thus present it in that spirit.

One of the many aspects of problem-solving is its psychological approach, which we will go into further detail in this chapter. The approach and the psychological set towards a problem situation likely affect the process and outcome of the challenge. Thus, we will explore these factors so that the cognizant teacher can anticipate difficulties students may have and provide appropriate remedies.

Other than its revelations in psychology, problem-solving can also be an excellent source and vehicle for enrichment in mathematics. Challenging problems, oftentimes those "off the beaten path", can lead to some very interesting investigations of mathematics outside of the regular curriculum, all while serving as wonderful opportunities to stretch the mind. Mathematics teachers must thereby, always be prepared to provide some appropriate material for the more gifted students, while simultaneously not neglecting the others, who also benefit from appropriately challenging problems. In this chapter we offer the math coach some ideas in this direction along with examples.

4.1. The Nature of Problem-Solving

Teachers build and develop problem-solving skills in concrete operational students by recognizing their inability to organize, systematize, and efficiently carry out the solutions, especially when several variables or relationships are involved. The approaches needed to compromise and bridge the gap between intuition and formal processes are those that help to organize data and relationships for systematic processing. In addition, students should be encouraged to use intelligent guessing and testing, or try any other strategy they may wish to use. In this way, more organized and efficient strategies can be assimilated naturally, built on the intuitive understanding and planning processes that the students can already recognize innately and use effectively.

Using example problems is the most straightforward tangible method for instruction in problem-solving. Two such approaches are possible.

1. The first approach is to select tasks that require the use and practice of specific methods.
2. The second approach is to select tasks that can elicit creative and insightful thought and consequently develop general problem-solving abilities.

There are many examples of mathematics problems in the normal secondary school course work; these can be used to develop creativity and insight.

The majority of such typical secondary school mathematics content can be taught with a problem-solving approach. To many teachers, this would be a rather new experience which requires much effort and additional planning. They may begin by having students work with problems whose solution leads to further investigations in the direction in which the course is aimed. This aroused curiosity then serves as motivation for further study.

Teachers who are otherwise inclined to pursue problem-solving for its own sake, rather than to fulfill the curriculum, may consider allowing students to work with pairs of problems that are similar in structure and that involve similar tasks. It is helpful for students to develop a good memory for problems and have experience with a variety of problem structures. Some useful activities could include the following:

1. Have students select from a second pair of problems that is similar to the first pair.
2. Ask students to write a pair of problems having the same structural relationships as the first pair.
3. Generalize the data and solution of the pairs.

Such exercises with problem recognition and ordering allow students to inspect the entire problem-solving process so that a form of metacognition can help develop a problem-solving sensitivity. This metacognition simply adds to the awareness of the need for and importance of problem-solving skills.

4.2. A Psychological View of Problem-Solving

In this chapter, we also consider the topic of problem-solving from the psychological point of view, with an aim to develop specific, and then easily recallable, problem-solving strategies. We will also inspect some example problems that contain interesting messages for further problem-solving and for a deeper understanding of mathematical concepts.

All problem-solving involves some form of information (perceptual, physiological, sensory) and the use of that information to reach a solution. Given the individual differences we find in development as well as the varieties of content and levels of complexity of problem situations, a single, simple approach to problem-solving would be difficult to discover (let alone implement).

As far back as 1910, John Dewey, in his book *How We Think* (Boston: D. C. Heath), outlined five steps for problem-solving. They were presented in the following order.

1. Recognizing that a problem exists — an awareness of a difficulty, a sense of frustration, wondering, or doubt.
2. Identifying the problem — clarification and definition, including designation of the goal to be sought, as defined by the situation that poses the problem.
3. Employing previous experiences, such as relevant information, former solutions, or ideas to formulate hypotheses and problem-solving propositions.
4. Testing, successively, hypotheses or possible solutions. If necessary, the problem may be reformulated.
5. Evaluating the solutions and drawing a conclusion based on the evidence. This involves incorporating the successful solution into one's existing understanding and applying it to other instances of the same problem.

Although not all problem-solving will necessarily follow this order, Dewey's analysis of the thinking process in problem-solving has not been improved on since. Note, that it involves both the intake or reception of information and discovery learning in an interrelated process — one in which the learner is an active participant in his own learning.

In terms of mathematical techniques, the work of George Polya in *How to Solve It* (Princeton University Press, 2004) presents instructions for problem-solving that not only are interesting but are also meant to ensure that principles learned in mathematics will transfer as widely as possible. His techniques are called *heuristics* (serving to discover), strategies that aid in solving problems. He says that there is a "grain of discovery" in the solution of any problem. "Your problem may be modest; but if it challenges your curiosity and brings into play your inventive facilities, and if you solve it by your own means, you may experience the tension and enjoy the triumph of discovery." Polya suggests the following set of heuristic methods.

1. *Understand the problem.* What is the unknown? What are the data? What is the condition? Draw a figure, introduce suitable notation. Separate the various parts of the condition.
2. *Devise a plan.* Find the connection between the data and the unknown. Have you seen it before? Do you know a related problem?
3. *Carry out the plan.* Check each step. Can you see that each step is correct? Can you prove that it is correct?
4. *Look back.* Examine the solution obtained. Can you check the result? Can you check the argument? Can you derive the result differently? Can you see it at a glance? Can you use a result or method for some other problem?

Perhaps the best way to support teachers to be effective problem solvers and teachers of problem solving is to provide them with many examples covering a variety of problem-solving techniques. Naturally, it is helpful at the beginning to classify the techniques shown, although over time it is wise for the math coach to work with teachers so as to enable students to discover the appropriate technique for the solution to a particular problem, as it is integral to the problem-solving process.

When a problem situation is presented, the raw information must be processed and refined to an acceptable answer. The processing is usually a multistep operation in which each step feeds off the results of the preceding one, using whatever available equipment is stored in the problem solver's arsenal of skills and knowledge. The first of these steps is the planning step, which is the development of a processing schedule, designed for the specific problem being considered.

Successful conception of a good plan is the primary achievement in the problem-solving process, but it is also the most difficult part of the process to teach. These metacognitive activities are tricky as they involve making connections between the intelligible parts of a problem statement, and the students' previous knowledge and experiences. This process continues until the problem can be classified into some already familiar group, ready for solution.

The ability to classify and to have groups of problem-types available is crucial to the process. Sometimes this may require breaking down a problem into smaller parts that are simpler to classify. Monitoring this process by the problem solver is essential as this self-awareness produces *control* of the processing, which is the key to successful problem-solving. It is perfectly normal, even desirable, to talk to yourself (sub-vocally) when working on a mathematics problem, as a way of trying to control the problem-solving process. This allows you to be in control and have a mastery of the necessary heuristics so you can select and pursue the correct approach for the solution.

There are a number of possible control decisions that ought to be considered:

- **Thoughtless decisions** move the process in scattered directions and do not build on any previous experiences or knowledge.
- **Impatient decisions** either stop the process entirely or keep the problem solver moving directionless in quest of a solution without even seeing a path to a conclusion, either successful or unsuccessful.
- **Constructive decisions** involve carefully monitoring control while employing knowledge and skills in a meaningful way, using proper solution paths and abandoning unsuccessful ones.
- **Immediate procedure decisions** require no control, since they simply access the appropriate solution path stored in long-term memory.
- **Non-decision** results when the statement of the problem is so perplexing that no knowledge or prior experience is helpful in the solution and the problem solver gives up.

An awareness of the problem-solving process is the first step for attaining control which enables the learner to find a proper solution path.

4.3. An Introduction to Problem-Solving Strategies

When students encounter problems in their everyday school lives, their response is not much different from those given to other personal experiences. These experiences can range from recognizing a problem as one very similar to one previously solved, to taking on a homework exercise reminiscent of those presented in class earlier that day. In such cases, the student is not doing any problem-solving, but rather is merely mimicking (or practicing) the earlier encountered situations. This repetitive behavior, seen in a vast majority of classrooms, is useful in attaining new skills. This holds true for acquiring problem-solving skills.

What needs to be made clear to teachers is that students who learn to consider problem-solving as an end in itself, but not merely as a means to an end, will benefit greatly in class as well as in their everyday lives. Problem-solving can be the vehicle used to introduce students to the beauty that is inherent in mathematics; it can be the unifying thread that ties their mathematics experiences together into a meaningful whole.

One immediate goal would be to have students become familiar with numerous problem-solving strategies and to practice using them. The results of this procedure would show itself in the way they approach problems and ultimately solve them. Enough practice of this kind should make a longer-range goal attainable, namely, that students would come to use these same problem-solving strategies in solving not only mathematical problems, but also in resolving problems in everyday life. This back and forth transfer of learning can be best realized by introducing problem-solving strategies in both mathematical and real-life situations concomitantly. Changing an instructional program by relinquishing some of its time-honored emphasis upon isolated topics and concepts, and devoting the time to a procedural approach, will require teacher support to succeed. Teachers must realize that the end results will prepare more capable students for this era, where the ability to think, rather than regurgitate a previously recognized solution, is becoming more and more important as technology advances.

Everyone seems to agree that problem-solving and reasoning are, and must be, an integral part of any good instructional program. Then why has this not come to pass? The major impediment to a successful problem-solving component in the regular school curriculum might be the neglected problem-solving training for teachers, as well as the lack of attention paid to the ways in which these skills can be smoothly incorporated into a regular teaching program. This is where the math coach should fill in the gaps. Teachers need to focus their attention on what problem-solving is, how they should utilize problem-solving, and how it should be presented to their students. Unfortunately, in recent years where teachers are being rated by their students' performance on standardized tests, they tend to emphasize "teaching to the test" thereby setting problem-solving skills aside. Problem-solving can be thought of in three different ways:

1. Problem-solving is a subject for study in and of itself.
2. Problem-solving is a way to approach a particular problem.
3. Problem-solving is a way of teaching.

Although all these ways are correct, the third concept is the overriding one for mathematics teachers to consider – problem-solving should become an integral part of their teaching process. At this point, we recommend three books that the coach might encourage teachers to consult:

1. *Problem-Solving Strategies for Efficient and Elegant Solutions: Grades 6 to 12* by A. S. Posamentier and S. Krulik (Corwin press, 2008).

2. *Problem-Solving Strategies in Mathematics: From Common Approaches to Exemplary Strategies,* by A. S. Posamentier and S. Krulik (World Scientific Publishing, 2015).
3. *Problem-Solving in Mathematics: Grades 3–6,* by A. S. Posamentier and S. Krulik (Corwin Press, 2009).

The math coach should be aware that teachers must initially focus their attention on their own ability to become problem solvers before trying to impart this skill. They must learn what problem-solving strategies are available to them, what they entail, and when and how they can be used. They must then learn to apply these strategies, not only to mathematical situations but also to everyday life experiences, so as to further convince students of the importance of the skill of problem solving. Often, simple problems can be used in clever ways to demonstrate these strategies. Challenging problems will naturally show the power of the problem-solving strategies. By progressively learning the strategies, beginning with simple applications of them and then moving toward solving more complex problems, students will have opportunities to grow in the everyday use of their problem-solving skills, which then can be further strengthened as they are used in more mathematical problems.

As a coach, you might want to immerse teachers in sessions where they familiarize themselves with the ten strategies that we will be listing later in this chapter, and to discuss with them how they can regularly incorporate them into their teaching. Often, by naming the strategies and highlighting their use during the instructional sessions, teachers will begin to use them as a tool as they will have a handle by which to grab when in need.

It is rare that a problem can be solved using all of the ten strategies, and it is equally rare that a single strategy can be used to solve a given problem. Rather, the utilization of a combination of strategies is the most likely occurrence. Thus, it is best to become familiar with all the strategies and to develop facility in using them when appropriate. Though the strategies selected here are not the only ones available, they represent those most applicable to mathematics instruction in the schools.

The user will thus, for the most part, determine suitability of a strategy in a particular problem. Determining the appropriate strategy is analogous to a repairman, who, when called upon to fix a problem, must decide which tool to use. The more tools he has available, the better we would expect his results to be. However, just as not every task the repairman has to do will be possible using the tools in his toolbox, not every mathematics problem will be solvable using the strategies presented here. In either case, experience and judgment play an important role. The coach's guidance will be quite useful here.

Every teacher, in order to help students learn and use the strategies of problem-solving, must have a collection from which to draw examples. The strategies that we will be discussing are labeled clearly so they can be called upon quickly as they are needed. In comparison, the repairman, in deciding which tool to use in repairing something, will usually refer to the tool by name.

All of the strategies presented can, and ought to, be regularly applied to everyday-life decision-making processes (or problem-solving in real-life situations). This practice should solidify their use and understanding and make their application to a mathematical context more natural.

To better aid in understanding the strategies presented here, we shall introduce each with a description, apply it to an everyday problem situation, and then present examples of how it can be applied in mathematics. In each case, the illustrations are not necessarily meant to be typical, but are presented merely to best illustrate the use of the particular strategy. Following are the strategies that we will consider:

1. Working backwards
2. Finding a pattern
3. Adopting a different point of view

4. Solving a simpler analogous problem (specification without loss of generality)
5. Considering extreme cases
6. Making a drawing (visual representation)
7. Intelligent guessing and testing (including approximation)
8. Accounting for all possibilities (exhaustive listing)
9. Organizing data
10. Logical reasoning

As already mentioned, there is hardly ever a unique way to solve a problem; some problems lend themselves to a wide variety of solution methods. As a rule, students should be encouraged to consider alternative solutions to a problem, such as by considering classmates' solutions and comparing them to the "standard" solution (that is, one given in the textbook or supplied by the teacher). Remember, many problems may require more than a single strategy for solution. The data given in the problem statement, rather than merely the nature of the problem, can determine the best strategy to be used in solving the problem. All aspects of a particular problem must be carefully inspected before embarking on a particular strategy.

4.4. The Ten Problem-Solving Strategies

We shall now consider these ten strategies along with mathematical examples for each that the coach can use to impress the value of the technique.

4.4.1. *Working backwards*

Although we use this problem-solving strategy quite often in everyday-life decision-making situations, it is not a natural method to call upon when tackling a mathematics problem. We use this method when developing a schedule for various tasks that must be completed by a certain time. We often start with what has to be done, the time at which all the work must be completed, and how long each task should take. We then work backwards to assign time slots to each task, and thus, to arrive at the appropriate time to begin the work.

The working-backwards strategy is also widely used every day in traffic investigations. When the police investigate an automobile accident, they must begin to work backwards from the time of the accident to see what the causes were, which car swerved immediately before the collision, who hit whom, which driver was at fault, what were the weather conditions at the time of the accident, and so on, as they attempt to reconstruct the accident.

When we look at the procedures which students are shown in many of their typical textbook exercises, we sometimes see very useful techniques presented. Unfortunately, these are often taken for granted and not called to the students' attention. Students may be required to reason in the reverse order, even though they have not been told to do so.

Example 4.1: An obvious example is the procedure students should use when writing proofs in the high school geometry course. They should begin by examining what they are trying to prove before doing anything else. Thus, an attempt to prove line segments congruent might stem from proving a pair of triangles congruent. This, in turn, should suggest that the students look for the parts necessary to reach this triangle congruence. Continuing in this manner, the students will be led to examine the given information. They are, in essence, *working backwards.*

When there is a unique end-point (that which is to be proved) and a variety of paths to get to the starting point, the working backwards strategy may be desirable. However, the *working-forward* method is still the most natural method for solving the problem. In fact, working in the forward direction is used to solve *most* problems. We are not saying that all problems should be attempted by the working-backwards strategy. Rather, after a natural approach (usually *forward*) has been examined, a backwards strategy might be considered to see if this provides a more efficient, more interesting, or more satisfying solution to the problem.

Example 4.2: Although many problems may require some reverse reasoning (even if only to a small extent), there are some problems whose solution is dramatically facilitated by working backwards. Consider the following problem, but beware it is not typical of the school curriculum, but rather a dramatic illustration of the power of working backwards.

> *The sum of two numbers is 12 and the product of the same two numbers is 4. Find the sum of the reciprocals of the two numbers.*

When students are faced with this problem the typical response is to write the two equations that are described in the first sentence, namely,

$$x + y = 12,$$
$$xy = 4.$$

Having been taught how to solve two equations simultaneously, they will proceed to do that and end up with a rather uncomfortable result:

$$x = 6 + 4\sqrt{2}, \quad \text{and} \quad y = 6 - 4\sqrt{2}.$$

They would then be required to find the reciprocals of these two numbers and then their sum. This is a rather cumbersome method to reach the solution. However, approaching this problem in reverse and asking to find the sum of the reciprocals,

$$\frac{1}{x} + \frac{1}{y} = \frac{x+y}{xy}$$

which then gives the answer directly through substituting in the numerator and denominator appropriately:

$$\frac{x+y}{xy}=\frac{12}{4}=3.$$

This should convince your teachers of the power of this working-backwards strategy.

Teachers should note that special care must be taken to avoid having students, discouraged by their frustrations, take the attitude that "I would never have been able to come up with that trick solution." Instead they should be encouraged to see that this valuable and unusual problem-solving strategy is one with which they should become familiar, and its use is indeed attainable with diligent practice.

4.4.2. *Finding a pattern*

One of the inherent beauties of mathematics is the logic and order that it exudes. This logic can be seen "physically" as a pattern, or as a series of patterns. The non-mathematician often appreciates geometry for the patterns it provides, but a mathematician uses patterns as an aid to solve problems not only in geometry, but in many other fields as well. We see mathematics problems in the secondary school curriculum that clearly require pattern recognition for their solution.

Example 4.3: To find the next two numbers in the sequence 1, 3, 4, 7, 11, 18, … , we must first search for and then recognize a pattern. One possible pattern has each number after the first two as the sum of the two preceding numbers. Recognizing that this is an example of a Fibonacci-type sequence (known as the Lucas numbers) leads to the next two numbers, namely 29 and 47. It is true that there may be more sophisticated (and more cumbersome) ways to find the immediate successors of 18 in the given sequence. However, consider finding the next two terms of the sequence 1, 10, 2, 7, 3, 4, 4, … . It is very unlikely that this can be resolved in any way other than by recognizing that this sequence is actually two separate sequences interwoven. One sequence is from the numbers in the odd positions: 1, 2, 3, and 4 (a difference of 1). The other is formed by the numbers in the even positions, namely, 10, 7, 4, with the difference of −3. Therefore, the next two terms would be 1 and 5.

These are problems whose solutions are, by their very nature, one of pattern recognition. In these cases, the use of pattern seeking was announced by the very type of problem. We will consider problems where pattern recognition is not expected, yet proves to be an invaluable aid to solving the given problem more simply than the traditional or common solution.

As coach, you should encourage teachers to consider that in everyday-life situations we (sometimes subconsciously) use pattern recognition to deal with a problem.

- For example, when you are looking for an even-numbered address on a street, you will look to the side where the even numbers are, and you will search for it in numerical order.
- The police often make use of the finding a pattern strategy in other ways, too. For example, when the police are confronted by a series of crimes (say robberies), they often try to find a pattern in the crimes as they look for a modus operandi that might lead them to a certain criminal.
- Scientists involved in medical research often utilize the finding a pattern strategy to permit them to locate and isolate similar variables, and thus draw conclusions about the particular virus or bacteria they are examining.

Example 4.4: To look for a pattern in a mathematics problem that calls for finding a pattern is not what this problem-solving technique is all about. The technique of finding a pattern is most useful when the problem does not call for a pattern to be found. Consider the following problem.

Find the sum of the first 20 odd numbers.

The problem calls for a simple addition, which is rather trivial to you. However, there is a clever procedure that can be used that involves recognizing a pattern. In this case, as we notice that the task is to find the sum:

$$1 + 3 + 5 + 7 + \ldots + 35 + 37 + 39,$$

we notice a pattern that adding the numbers in pairs:

$$1 + 39,\ 3 + 37,\ 5 + 35,\ \ldots .$$

Since we had 10 pairs of numbers each of which has a sum of 40, we can get a sum of the sequence as 400. This is simply an illustration that one can show teachers, which they can then take further to their classes to demonstrate the advantage of looking for pattern. We can examine this problem by *looking for a pattern,* but in a different manner.

Addends	**Number of Addends**	**Sums**
1	1	$1 = 1^2$
$1 + 3$	2	$4 = 2^2$
$1 + 3 + 5$	3	$9 = 3^2$
$1 + 3 + 5 + 7$	4	$16 = 4^2$
$\vdots$	$\vdots$	$\vdots$
$1 + 3 + 5 + 7 + \ldots + 37$	19	$361 = 19^2$
$1 + 3 + 5 + 7 + \ldots + 37 + 39$	20	$400 = 20^2$

4.4.3. *Adopting a different point of view*

This is a very useful strategy that your teachers might find an opportunity to make a dramatic difference in a solution. It requires "forcing" yourself to attempt solving a problem by thinking about it in a contrasting way.

Example 4.5: An example problem is as follows:

> *For a 25-team league, find the number of games that must be played to determine a champion in a single-elimination tournament.*

Most students faced with this problem will probably try to simulate the situation and begin with 12 teams eliminated after the first round (requiring 12 games played), and then continue considering only the winners, leaving the losers eliminated. This approach should lead to a correct answer, although it is a bit tedious. By adopting a different point of view, one might consider counting the *losers*. How many losers must there be? There must be 24 losers to get a champion. The number of games played to get 24 losers is 24. So, the problem has been simply solved just by taking a different point of view.

When introducing this to teachers, you might give them an example such as the following, where they can see the strategy is beyond mathematics. In an everyday-life discussion with a friend, it sometimes helps to consider the friend's point of view as a way to resolving an issue. In a debate, this is a particularly useful strategy. Another illustration of how this technique can be useful is in taking attendance in a class. Rather than to call out the names of those present, consider adopting another point of view and count the absentees. The rest are then present.

4.4.4. *Solving a simpler analogous problem*

This method is one that teachers should be made aware of to be used more frequently, as it turns out to be most revealing. The essence of this technique is to change the given problem into an easier ancillary problem and, after solving it, gain the insight needed to solve the original problem. This particular strategy, *solving a simpler analogous problem,* can be referred to as specification without loss of generality. That is, if no restrictions are given in the problem, we may select a special case of the given situation to examine.

Oftentimes, this strategy may be used to make a problem easier to grasp by replacing some numbers or variables with easier ones and then reverting to the original problem, as sometimes a problem can appear to be unusually overwhelming. Yet by considering a simpler case of the situation presented, the problem can become much more manageable.

Example 4.6: Take for example the following problem:

> *Given that the sum of vertex angles of all pentagrams (five cornered stars) is constant, what is the sum of these vertex angles:* $\angle A + \angle B + \angle C + \angle D + \angle E$ (See Figure 4.1).

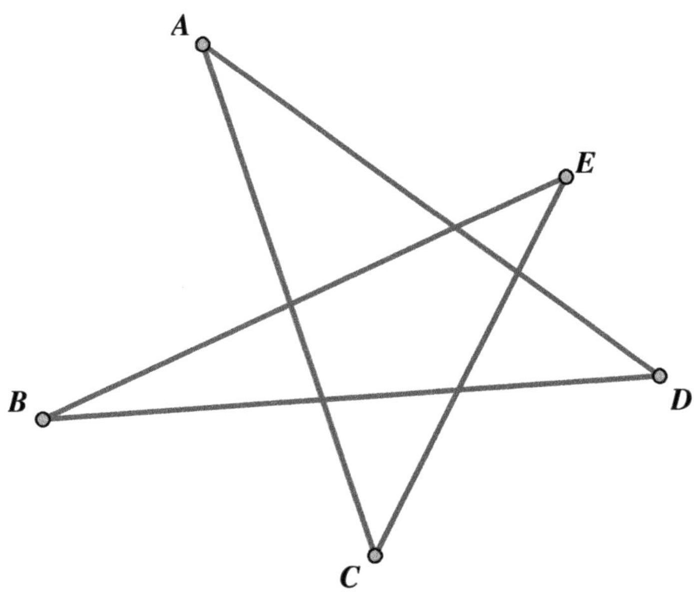

Figure 4.1

Since the type of pentagram was not specified, we can assume that the pentagram is either a regular pentagram or one that can be inscribed in a circle, where all the vertices lie on the circle as shown

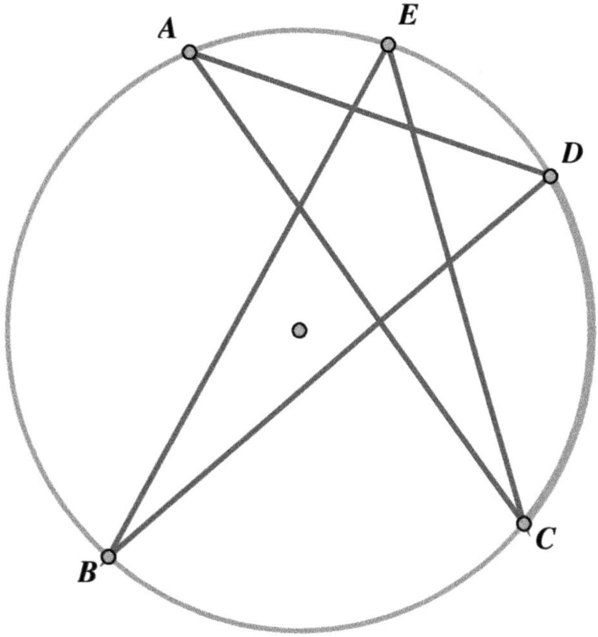

Figure 4.2

in Figure 4.2. In that case, each of the angles is measured by one half of respective intercepted arc. For example, angle A is one half the measure of arc DC. And so, if each of the angles is one half of their respective intercepted arcs, and the total measure of these intercepted arcs is 360°, then the sum of the angles would be one half of that, namely, 180°.

Another special case which could be used to demonstrate this strategy would be to assume the pentagram is a regular one and then find the angle sum (where each of the angles has the same measure).

4.4.5. *Considering extreme cases*

This is a method that your teachers should probably be able to relate to and may not even be aware of as a clever strategy for mathematics problem-solving. To analyze some situations, whether in a mathematical setting or not, it can be helpful to look at extreme cases. Holding some variables constant, while varying others to extremes, sometimes yields useful insights, which facilitate solving some problems with ease. By considering extremes, we may be changing variables in the question, but only those that do not affect the crucial variables of the problem. However, we must be careful not to change a variable that affects other variables.

Used properly, considering extreme cases can be one of the most useful strategies for solving mathematical problems as well as those in everyday life, such as a situation in the context of what the "worst case scenario" would be. That is, what is the worst that can happen if our argument goes awry? Then we proceed. Anticipating a worst-case scenario is a form of extreme case consideration. We use it frequently in everyday life when budgeting time, budgeting finances, and so on.

Example 4.7: To best grasp the value of this technique, consider the following problem. Have your teachers/students attempt to solve it by working together in groups.

> *Two concentric circles are 20 units apart as shown in Figure 4.3. What is the difference between the circumferences of the circle?*

As coach you might want to encourage teachers to guide students who attempt this problem ina traditional straightforward solution, which is to find the diameters of the two circles and then in turn find their circumferences and subtract them (see Figure 4.3). You, as the math coach, might want to guide them to this traditional solution by letting the diameter of the smaller circle be represented by d, then the larger circle diameter would be $d + 40$. It follows then that the respective circumferences are πd and $\pi(d + 40)$, and the difference is 40π.

Nevertheless, a more elegant procedure would be to use an extreme case. Let the smaller of the two circles become smaller and

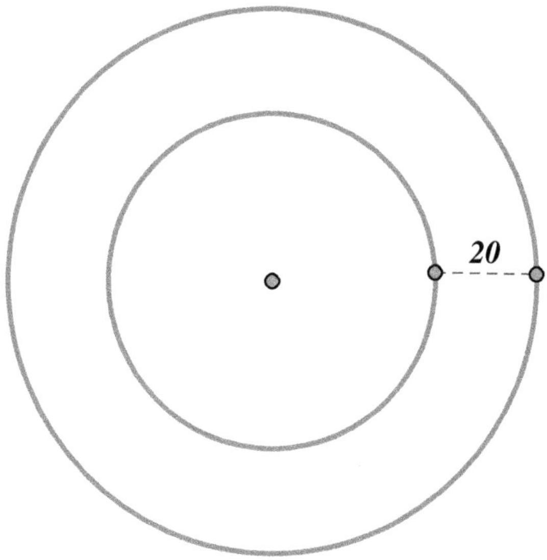

Figure 4.3

smaller, until it reaches an "extreme" and becomes a point, which is the center of the larger circle. Now the difference between the circumferences is merely a circumference of the larger circle whose diameter is 40, and who circumference is then 40π. Teachers should recognize the power and beauty of this procedure of considering extremes.

4.4.6. *Making a drawing (visual representation)*

It is expected of teachers to already know how to make drawings to solve geometric problems; not to do so is foolhardy! However, for this strategy we are considering the use of a drawing to solve a problem, where a visual representation is not the usual approach based on the nature of the problem. In real life, there are many decisions which are made based on the visual presentation of data and relationships, where a pictorial representation acts more as a facilitator than as an expected element of the situation. Drawing a picture often makes the description clearer and easier to follow. After all, it has been said many times that one picture is worth a thousand words!

Example 4.8: Consider the following problem, one in which it is not expected that a diagram should be drawn.

> *At 5 o'clock, a clock strikes 5 chimes in 5 seconds. How long will it take the same clock to strike 10 chimes (at the same rate) at 10 o'clock?*

We hope that experienced math teachers will immediately realize that the answer is not 10 seconds! While there is no indication in this problem that a diagram would be of help, let us nonetheless use a drawing of this situation to see exactly what is taking place. In Figure 4.4, we will let each dot represent one chime of the clock. Then the explanation next to the diagram tells the whole story and the problem is easily solved. You might suggest that teachers search for similar verbal problems, which lend themselves to a diagram, where one is not intuitively expected.

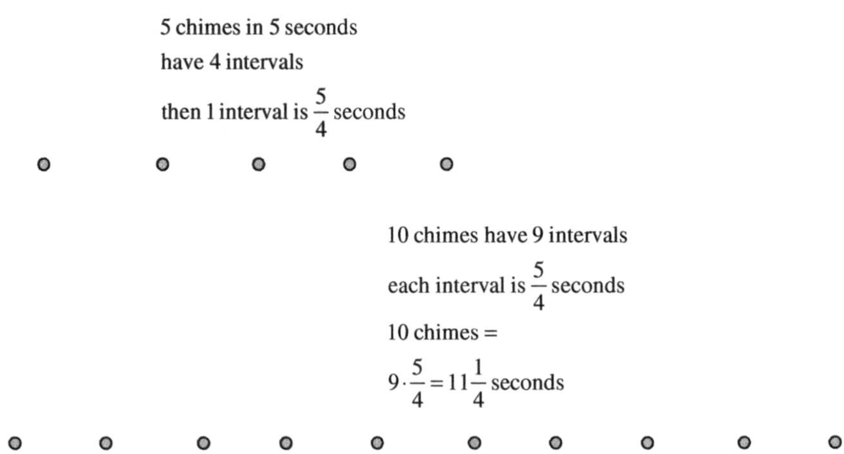

Figure 4.4

4.4.7. *Intelligent guessing, testing, and approximation*

This technique is often referred to as the method of "trial-and-error", and most teachers are probably familiar with this process, but have been prone to disregard its mathematics context as it is not necessarily seen as a popular problem-solving procedure. But that name is an oversimplification, for this problem-solving strategy is, in its own right, quite sophisticated.

Intelligent guessing and testing is particularly useful when we need to limit the values for a variable to make the solution more manageable. It is also helpful when the general case may be far more complicated than a specific case. By approximating, we can try to narrow down the options in an effort to focus on the correct answer. By the way, using this strategy employs thinking about the problem beyond a mechanical procedure, which is something often neglected in the secondary school math curriculum. In using this strategy, we make a guess, then test it against the conditions of the problem, with each successive guess based upon information obtained from testing the previous. Keep in mind that there is a great deal of difference between a "guess" and an "intelligent guess."

Solving an equation is really only a variation of intelligent guessing and testing. What we are doing in the "solving" part is putting forth a guess, and eventually arriving at a solution intelligently by some careful mathematical manipulations. It is in the checking or verification part that we are testing our guess to show that it is, indeed, correct.

Example 4.9: To demonstrate that there could be a place for this strategy in our problem-solving arsenal, let's consider a convincing example that you might want to show your fellow teachers. Were someone to ask the question "*Find three consecutive whole numbers whose product is 24*", the algebraic solution could be quite challenging. The equation would look like this:

$$(x)(x+1)(x+2) = 24,$$

which is a cubic equation and not too easily solved. However, using the *intelligent guessing and testing* strategy, we can find the three numbers easily: 2, 3, and 4. Another more realistic example of this strategy is found in the solution of the following problem.

> *Find two positive integers that differ by 5, and the sum of their square roots is also 5.*

At first sight, the problem seems rather straightforward and perhaps easily solved algebraically as follows:

$$x - y = 5, \text{ and } \sqrt{x} - \sqrt{y} = 5$$

Substituting:

$$\sqrt{x} - \sqrt{x+5} = 5$$

Squaring both sides:

$$2\sqrt{x(x+5)} = -2x + 20$$

$$\text{then } x = 4, \text{ and } y = 9$$

Most teachers will balk at this solution process, which opens the door to an alternative strategy namely, *intelligent guessing and testing.* Here is the thinking that would go into this strategy: since the sum of the square roots of the two integers is five, the individual square roots must be 1 and 4, or 3 and 2. This would indicate that the integers are either 16 and 1, or 9 and 4. However, only the second pair has a difference of 5 and so that must be the correct answer. It would be a good practice for teachers to search for other illustrations of where this strategy can be dramatically demonstrated.

4.4.8. *Accounting for all possibilities*

Once again, we are proposing a strategy that some teachers may not be too comfortable using, since it is not necessarily a strategy that is taught in the classroom. Although there may be instances where this strategy is not the most sophisticated procedure, it may produce the simplest and most effective solution, since it is typically not very abstract. However, the necessity of accounting for *all* the possibilities is crucial in the use of this strategy. If one does not have an organized procedure for accounting for all the scenarios, the strategy often goes awry.

In everyday life, we tend to use this procedure without even being aware of it. For example, when a lamp ceases to work, we consider all possibilities for the malfunction. We list such things as a fault in the electric wire, a burned-out bulb, and that circuit breaker, or a faulty outlet, etc. until we find the true cause.

Example 4.10: In mathematics, there are many examples where the preferred strategy would be to account for all possibilities. To demonstrate this, we suggest the following problem:

If 4 coins are tossed, what is the probability that at least two heads will be showing?

Naturally, your teachers will reply that typical probability methods will generate the correct answer. However, it is quite simple to list all possibilities of the sample space and then highlight (bold type) those that meet the requirements — that there are at least two heads showing (see Figure 4.5).

HHHH	**HHHT**	**HHTH**	**HTHH**
THHH	**HHTT**	**HTHT**	**THHT**
HTTH	**THTH**	**TTHH**	HTTT
THTT	TTHT	TTTH	TTTT

Figure 4.5

The bold events are those that have two or more H's and satisfy the given conditions. There are 11 of these; thus, the required probability is $\frac{11}{16}$.

<u>**Example 4.11**</u>: This illustration may not convince your teachers of the power strategy, so we offer another application of this strategy as follows:

> *In triangle ABC,* $\cos \angle A \cdot \cos \angle B \cdot \cos \angle C > 0$. *What kind of a triangle is* ΔABC*?*

We believe that your fellow teachers will realize the inherent difficulties in trying to solve this inequality algebraically. This will conveniently open the door for a strategy that will *consider all possibilities* of types of triangles. We consider the following three possibilities:

1. *Triangle ABC is a right triangle.*
If triangle ABC is a right triangle, then one of its angles must measure 90°, and $\cos 90° = 0$. Then the above product would be equal to 0, thereby not allowing us to conclude that this is a right triangle.

2. *Triangle ABC is an obtuse triangle.*
If triangle ABC is an obtuse triangle, then one of its angles (let us assume it is angle B) must have measure greater than 90°, while angle A and angle C must both be acute. If this is the case, then $\cos \angle B < 0$, while $\cos \angle A > 0$, and $\cos \angle C > 0$. Therefore, the product is once again less than 0, and doesn't meet the requested criterion.

3. *Triangle ABC is an acute triangle*
If triangle ABC is acute, then all of its three angles must be acute. In that case the cosine function of each angle is greater than 0 and

so the product is also greater than 0. Thus, the triangle is an acute triangle.

4.4.9. *Organizing data*

Once again, the coach's task is to try to convince teachers that there are alternative ways to solve problems and methods that not necessarily conform to the textbook situation of a "mechanical solution," where the student is required to do little thinking. It is not uncommon to find a student who, a bit baffled by a problem, emerges from this state of confusion by organizing given data from the problem situation in a way different from the way it was presented. This reorganization may be visual, or it may simply be an alternate way to look at the situation, thus, helping one analyze concepts. Just such a question could appear on a test, so that the student who has the ability to first organize data and ideas and then analyze them is at a distinct advantage.

Example 4.12: Organizing data in the solution to a mathematics problem can manifest itself in various ways, and one such way can be seen in the following problem's solution. To convince your teachers of the power of this strategy, we offer the following problem:

> *Find the greatest possible product of two natural numbers whose sum is 41.*

The typical solution has students setting up the following equation:

$$y = x(41 - x),$$

which is a parabola with a turning point that would give the maximum value. However, we can easily solve the problem by *organizing the data* in tabular form:

Numbers	Product
1×40	40
2×39	78
3×38	114
⋮	⋮
15×26	390
16×25	400
17×24	408
18×23	414
19×22	418
20×21	420

Thus, the greatest possible product is 420.

Example 4.13: Another form of *organizing the data* can be seen with the solution to the following problem.

> *If **A** apples cost **D** dollars, what is the cost in cents of **B** apples at the same rate.*

Although there are lots of approaches that teachers will show students, yet again one of the more effective ways is to organize the data in a meaningful fashion.

As a general rule, a problem like this one can best be solved by *organizing the data* in some useful manner. Here, we will use proportionality along with some common sense. The proportionality is obtained by setting up the equation with the same units of measure in each fraction:

$$\frac{A}{B} = \frac{\text{cost of } A \text{ Apples}}{\text{cost of } B \text{ Apples}} = \frac{100D}{x}.$$

Notice that we have used common sense in obtaining the last fraction. Since the problem called for the answer in cents, we use a fraction with cents as the unit rather than dollars. Thus, when we find x, we have found the answer. The rest is simple:

$$x = \frac{100BD}{A},$$

It should be pointed out to teachers that there needs to be a concentrated effort in having students learn to think proportionally. The weakness shown by many students with a problem of this sort results from the neglect in many curricula, which do not properly stress proportional thinking. This can also be seen with adults, who may have not been trained to think in this fashion.

4.4.10. *Logical reasoning*

As a math coach you will sometimes have to convince teachers that there is life beyond the mathematics class. Effortlessly, lessons drawn from such events are easily carried over to the classroom or any other conversation. For example, when dealing with friends and colleagues, we find that what we say will often evoke a certain response; that response can then lead to another, and so on. When we try to predict a conversation scenario, or a potential discussion/argument, we are, in effect, using *logical reasoning*.

For instance, if you say A, then you expect the response will be B, which will lead to statement C, which will likely be responded to with statement D. Logical reasoning, done effectively, can improve interpersonal relationships by helping avoid problems before they even arise. We often analyze these situations without being aware of the actual process. However, in mathematics, we must consciously guide our students to be aware of this thinking process — to think logically. Since it can be argued that inductive thinking (i.e., going from several specific examples to a generalization) may be more natural, the logical form of reasoning requires some practice.

In everyday-life situations, we typically rely on logical reasoning to craft a strategy for a work plan, or we may use it to argue a point with a colleague or boss. The strengths of such arguments are often dependent on the impeccability of the logical reasoning used. How an argument is posed, and its validity can mean the difference between success and failure of an argument, an advancement on the job, and even status. In addition, the outcome of a business deal can depend on one's facility with logical reasoning. Almost every mathematics problem we tackle involves a degree of logical reasoning, even the selection of the most efficient problem-solving strategy.

Example 4.14: Formal logic is, hence, fundamental to pure mathematics and to proofs. Often, logical reasoning that does not seem to be a proof will enable analysis of a problem. When it is appropriate for students to craft proofs, it is suggested that they be given "prove-or-disprove" problems, often enough for them to develop the habit of trying out a conjecture before attempting to prove it. This approach is second nature to mathematicians confronting unfamiliar conjectures. Let's examine a problem that allows us to demonstrate this technique. Consider the following:

Find all pairs of prime numbers whose sum equals 999.

It might be surprising that some experienced teachers will initially wonder what this problem is driving at. Many students will begin by taking a list of prime numbers and trying various pairs to see if they can get 999 as a sum. This is very time-consuming and rather tedious and could possibly lead to frustration. Let's use our *logical reasoning* strategy to solve this problem. In order to obtain an odd sum for two numbers (prime or otherwise), exactly one of the numbers must be even. Since there is only one even prime, namely *2*, there must be only one pair of primes whose sum is *999*, and that pair is *2* and *997*.

4.4.11. *Summary of problem-solving strategies*

These ten problem-solving strategies must be practiced with motivating problems that show their power. The names of the strategies should be used whenever they are used, since research shows name repetition improves retention of the strategy. We suggest referring to the following book for lots of examples for each of these ten strategies, and the reinforcement necessary to become more facile with these techniques:

> Posamentier, A. S. and S. Krulik. *Problem-Solving Strategies for Efficient and Elegant Solutions: A Resource for the Mathematics Teacher.* Thousand Oaks, CA: Corwin Press, 2008.

Once you have familiarized your fellow teachers with the strategies, they should try to find ways to apply them to textbook problems and exercises, thereby reinforcing their use. Also, they should be encouraged to use the strategies to find alternative methods of solution, since how one gets to a solution is often as important as getting the solution. Remember, there is usually a "poet's way" — the elegant method, and a "peasant's way" — the inelegant method, of getting a solution. We would like all our students to eventually use the poet's way!

4.5. Creating Mathematical Problems

As the math coach, you might wish to discuss with the teachers you work with on how students can be trained to develop and solve their own self-made problems by making changes, however minute, in existing problems. When students develop their own problems, they may occasionally create problems that are beyond their ability or capacity. Some may even be unsolvable; however, no problem should avoid discussion.

Problems posed by students can have some conditions changed to produce a new problem or variation of the original problem. Thus, the teacher should determine some alternative assessment scheme for giving credit to students not only for solving problems correctly, as is traditionally the case, but also for creating problems. Through this creative effort, students really begin to understand what problems are all about. As an aid to teachers, the math coach may offer suggestions on how specific problems can be changed to produce new ones, demonstrating to students the types of variations that might be made. The real creativity, however, must come from their own variations of problems.

Example 4.15: Consider the following problem:

> *Charlie has 45 coins, consisting of nickels (5 cents) and dimes (10 cents). The total value of the coins is $3.50. How many coins of each kind does Charlie have?*

At this point it would be wise for teachers to consider possible variations of this problem, and we offer just a few here:

1. What is the largest number of coins in nickels and dimes he could have that would total $3.50?
2. What is the smallest number?
3. How would the problem change if it included quarters (25 cents) as well as nickels and dimes?
4. Is it possible to have only dimes and quarters instead of nickels and dimes?
5. In how many ways could $3.50 be represented in nickels and dimes?

Although this exercise will be seen by some teachers as a rather extraneous activity, as a math coach, you might have teachers consider that they will make the greatest impression on their students when they deviate cleverly from the typical classroom style. It is often felt that taking time out to stress problem-solving strategies takes valuable time away from working through the curriculum. Many feel that this is a misconception. By focusing on problem-solving, we are training our students to develop a skill that transcends mathematics and reaches into many other fields of study as well as everyday lives. It is also an investment, as we will probably be developing more effective learners, making the mathematics instruction more effective and more pleasant.

4.6. Using a Problem Deck

Let us make one last suggestion for helping teachers teach in a problem-solving mode. The *problem deck* provides an excellent resource, and it can be produced in either of two formats (or both, if desired).

4.6.1. *Student deck*

Created by coaches/teachers for students: Using a pack of 5″ × 9″ file cards, write a problem that would challenge the students on the front. On the back, prepare a set of questions and/or hints for the students. Place the deck in a resource corner or lab center. As a student completes his or her work, have them select and attempt to solve one of the problem cards, while keeping track of which problems prove to be most popular. Using this, constantly add to the deck as you discover new problems within the students' proficiency. Make it a point to always discuss with students their solution in a one-on-one setting.

4.6.2. *Teacher deck*

Created by coaches for teachers: Using the same cards, write a problem appropriate to your class level on the front. Also, indicate where in the curriculum the problem fits. On the back, place one or more solutions, the answer, and some hints for resolving the problem situation. The deck provides an excellent resource for teachers and should continue to grow with problem-solving experience. Make sure to keep track of which problems are too difficult, and which are too easy. As the deck grows, some teachers may put problem cards in their plan book for use in their lessons.

Encourage other teachers to share problems where appropriate, while ensuring the problems are adjusted to match the grade level and abilities of the students. As the decks grow, the presented problems become more and more integral to the teachers' lessons.

Chapter 5

Enriching Mathematics Instruction

In the education community, we talk about enriching the instructional program that is designated by a standard curriculum. This is clearly a mantra that a coach must carry to the teachers. Typically, enrichment is seen as a very worthwhile venture as it is an excellent way to motivate students for the subject matter, something that is very necessary in today's world, especially with the importance of mathematics instruction for the technological era. Most standards either suggest or specify that enrichment should be an integral part of the instructional program. However, this suggestion rarely makes it to the classroom, as some teachers lack the resources to identify the appropriate items to use to enrich their instruction. Here is one point where the coach can be very helpful in identifying methods of enrichment. There are also some teachers who incorrectly believe that time is wasted on enrichment or on topics which will not be tested. This is an incorrect notion, since we believe that by enriching the instruction and thereby, motivating students to love mathematics, an investment is made where the time spent will gain a much more enthusiastic and receptive learner. As a math coach, you should see how this can enhance the instructional process. Therefore, it is incumbent upon the coach to carry this message forward to teachers, telling them to spend a few minutes periodically to win the students over to be strong supporters of mathematics learning.

First, the coach must emphatically state that mathematics enrichment is *not* reserved for gifted youngsters — students of all ability levels should have their mathematics instruction enriched. In this chapter, we shall explore ways in which a coach might work with teachers to encourage them to enrich mathematics instruction for all students.

Enrichment of mathematics can manifest itself in at least **three** different ways. The easiest, and least creative to implement, is ***acceleration***. This refers to the process of moving a better student along the mathematics path more rapidly. Although effective, acceleration has a few drawbacks. First, a student can be moved along so fast that he or she accelerates right out of the mathematics courses that are offered at the school. That is, the student may finish the coursework a high school has to offer, and then be left with no further mathematics courses to take, while not having completed the other subjects of the high school coursework. Albeit down the road, the math coach needs to bear in mind that to have a hiatus of a semester or more in high school without any mathematics instruction is counterproductive, in that the student will have lost the continuity of the development of the subject matter. When this is the case, the math coach should arrange for a student to work privately with a teacher who volunteers to be a mentor for the remainder of the student's high school career, or the math coach may seek out opportunities for the student to enroll in a course at a nearby college (if that option exists), or the worst option would be that the student "takes a vacation" from mathematics, which we indicated above is highly discouraged. This last option would be an inexcusable shame, for a bright student might ultimately be lost to mathematics study later on.

Another reason for enriching a bright student in ways other than to *accelerate* him or her is that such other forms of enrichment, which we will recommend further along in this chapter, can stimulate the student to pursue mathematics study more seriously later on, or it may simply motivate the student to improve his or her understanding of mathematical concepts and ideas. There are many topics for a student to investigate — topics outside the regular school curriculum. As math coach it would be good to develop a bank of these topics, which might be considered "off the beaten path."

We have available another form of enrichment, which we may call ***expansion***. This term leverages on the resourcefulness of the mathematics teacher to delve into regular curriculum topics in greater detail than normally required. Presenting more trivia will better spur students on to investigate in greater depth topics regularly taught in the classroom. Once again, the math coach could be a resource for such ideas, as demonstrated when we revisit this in greater detail for gifted students later in this chapter.

Digression is another form of enrichment and it refers to taking class time to consider a topic that is not in the regular curriculum but relates to one that is. Moving out of the confines of the syllabus to related topics and studying them in appropriate detail can open up new areas of interest for students. One must keep in mind that this technique is more suitable for the more senior students, as younger students on average do not have the very sophisticated ability to abstract as do older, more mature students. Once again, we will consider digression for gifted students later on.

As mentioned earlier, the time taken away from syllabus topics to do enrichment activities must not be seen as time lost from the regular instruction. It is quite the contrary! The time used for enrichment activities is actually an investment of time, where students will become more active participants in the classroom and more efficient learners when their instructors spend time to motivate and interest them in mathematics. A "turned on" audience needs less time to learn and absorb new concepts. This is an essential part of the mission and responsibility of a mathematics coach.

Another form of enriching mathematics instruction is done with a historical approach. It is a popular myth that a mathematics class is lifeless and dull. Unfortunately, that myth is too often true — though it need not be so! Of course, as a math coach, you need to dispel that thinking. Undoubtedly, you may frequently find some teachers concentrating on the teaching of mathematics to reach a deadline, such as administering a test or completing a course of study. The luxury of teaching what mathematics is all about sometimes seems to be beyond their allotted time and thus, grasp. But is it really? As coach, you might encourage the teachers that they can easily teach where mathematics comes from, who first thought of it, and who later developed and refined it. In short, we can use the history of the subject, including the lives, loves, successes, and failures of the people who created it, to breathe life into what might otherwise be rather dull. There are times when history should be integrated into the subject matter in a light and lively way. We recommend a book by A. S. Posamentier and C. Spreitzer, *Math Makers* (Prometheus Books, 2019), which presents brief biographies of 50 famous mathematicians along with some of the key discoveries that each has provided.

For example, teachers could be shown how the original development of geometry in Egypt and Babylonia was a result of the desire of priests to construct temples and of kings to survey land for tax purposes. The techniques were crude and relied heavily on intuition but sufficiently accurate for their needs. We have evidence of some of these techniques in the Ahmes Papyrus, which was written about 1650 B.C.E. and discovered in the Nineteenth Century. Portions of it are now preserved in museums in London and New York. The papyrus contains formulas for calculating areas of rectangles, right triangles, and trapezoids that have one leg perpendicular to the bases, and for approximating the area of a circle. The Egyptians apparently developed these formulas from their experience with area of land.

The first mathematician who appeared to be dissatisfied with methods based solely on experience was Thales (ca. 640–546 B.C.E.). We honor him today as the man who always said, "Prove it!" And he frequently did. Among the better-known theorems first proved by Thales are:

- The base angles of an isosceles triangle are congruent.
- Vertical angles are congruent.
- An angle inscribed in a semicircle is a right angle.

These are all very simple things that can be shared with students to give them a prospective of where these concepts evolved from. Once again, it is incumbent upon the math coach to raise awareness in teachers of relevant historical backgrounds in mathematics.

Taking this another step further, consider Pythagoras (ca. 582–507 B.C.E.) and his followers, the Order of the Pythagoreans, who followed in the footsteps of Thales. They used his method of proof to develop not only the Pythagorean theorem, but also theorems concerning the sum of the measures of the angles of a polygon, the properties of parallel lines, the five regular solids, and incommensurable quantities.

Even though it might be a bit premature in the development of young students, teachers may carefully explain where their studies of mathematics will be going and, in that sense, it would be good to discuss the kind of thinking that we use in mathematics today. Teachers you work with might already be aware that it was Thales' work that marked the beginning of an era of mathematical development, in which deductive proof became the accepted method of logical reasoning. This method is used to derive theorems from postulates and, in this way, to develop a system of logically arranged statements. This era reached its apex with Euclid's *Elements,* 300 years after Thales. In the *Elements,* Euclid unified the work of the scholars who had preceded him by presenting all the mathematics known in his day in a systematic manner — a truly stupendous achievement. Much of his work was original, for by means of the deductive method he demonstrated the vast amount of knowledge that can be acquired through reasoning alone. Euclid included algebra, number theory, as well as geometry in his writings. The *Elements* turned out to be a work of major importance in the history of the civilized world. In later years it was translated from Greek to Arabic (ca. 800) and from Arabic to Latin (ca. 1120). The first printed edition appeared in Latin (1482), and many other editions followed. Next to the Bible, the *Elements* has been published in more editions and in more languages than any other book.

The original version of the *Elements* was written in thirteen separate parchment scrolls, or "books". The fifth theorem in Book I is the familiar "Base angles of an isosceles triangle are congruent." (*Isosceles* is derived from the Greek words *isos,* meaning equal, and *skelos,* meaning legs.) These are just some of the sorts of things that you might encourage teachers to tell their students as a means of digression and simultaneously interest them to feel that mathematics also has a history that can be exciting. If you wish to take it further, you might mention that throughout the early years, a test for students' cleverness was to measure their ability to prove that the base angles of an isosceles triangle are congruent. This was known as the "*pons asinorum*", or "*bridge of asses (fools)*" in the Middle Ages. The implication was that certain students had difficulty "crossing" this bridge in order to proceed further in their study of geometry.

If you would like teachers to talk about the development of mathematics, they could mention that throughout the course of history, but especially during the last four centuries, Euclid's *Elements* has endured unimaginable "torture". It has been simplified, complicated, projected, distorted, deformed, and sometimes changed beyond recognition. The result? A number of other approaches to the study of geometry, such as analytic geometry, projective geometry, topology, non-Euclidean geometry, logic, etc.

If any of teachers you work with are involved in geometric constructions using straightedge and compasses, then they might want to discuss some of the discoveries in that realm. For example, there were such constructions that eluded the capabilities of the best mathematicians for centuries. While certain polygons could be constructed with ease using the traditional tools of geometry, others, such as the regular seven-sided polygon (heptagon, or 7-gon) or regular 17-gon (or heptadecagon), seemed to be beyond their reach. In 1796, a 19-year-old, Carl Friedrich Gauss, who later became one of Germany's most famous mathematicians, proved that it was possible to construct a regular 17-sided polygon using only compasses and straightedge, and that certain regular polygons, such as the regular 7-gon, could not be so constructed. You might also want to mention in detail that what Gauss proved is that a regular polygon with an odd number of sides can be constructed if either of the following conditions is satisfied:

> **Condition 1**: The number of sides is equal to $2^{2^n}+1$, where n is a whole number, and $2^{2^n}+1$ is a prime number.
> **Condition 2**: The number of sides is the product of two or more different numbers obtained in Condition 1. (For example, you can construct a regular polygon of, say, fifteen sides, because $15 = 5 \times 3$, and 5 and 3 are numbers derived from the rule in Condition 1.)

The discovery of this remarkable theorem convinced Gauss to enter the field of mathematics for his life's work, instead of linguistics, in which he also excelled. A monument erected to him in Brunswick, Germany, at the place of his birth, consists merely of a regular 17-gon, the symbol of his great achievement.

Although algebra is traditionally taught at the secondary school level, it has become more common to be introduced at the elementary school level, and so a math coach should be aware of its broad aspects. The treatment of algebra by the Greeks was geometric. Much of the algebraic knowledge of the Greeks was brought together by Diophantus (ca. 275), a prominent mathematician, whose work on integral solutions to certain types of equations is still studied today. Incidentally, the first woman mathematician we hear of in ancient times is Hypatia (ca. 410), who wrote commentaries on the work of Diophantus.

Just to give the math coach more ideas, consider that in another part of the world, Hindu mathematicians were also concerned with methods for solving equations, especially the quadratic. One of their more interesting contributions to the subject is that they enjoyed posing problems in a colorful way. Here is one example (and many more can be found in books on the history of algebra). Consider this:

Of a collection of mango fruits, the king took $\frac{1}{6}$, *the queen took* $\frac{1}{5}$ *of the remainder, and the three chief princes each took* $\frac{1}{4}$, $\frac{1}{3}$, and $\frac{1}{7}$ *from that same remainder. The youngest child then took the last 31 mangoes. How many mangoes were in the original collection?*

This is a problem that the coach could give some teachers to get them warmed up to some simple algebra that is now being offered at the upper elementary level in some schools. For those seeking a model solution, one such is as follows:

Let T represent the total number of mangoes. The king took $\frac{1}{6}$ of T, leaving $\frac{5}{6}$ of T as the remainder. We shall call this remainder R.

Of R, the queen took $\frac{1}{5}$, and the princes each took $\frac{1}{4}$, $\frac{1}{3}$, and $\frac{1}{7}$, leaving 31 for the youngest child.

As a fraction of R, the youngest child must have taken:

$$R-\frac{1}{5}R-\frac{1}{4}R-\frac{1}{3}R-\frac{1}{7}R$$

Simplifying, and substituting a common denominator, the youngest child took:

$$R\left(1-\frac{84}{420}-\frac{105}{420}-\frac{140}{420}-\frac{60}{420}\right)$$

$$=R\left(1-\frac{389}{420}\right)$$

$$=\frac{31}{420}R$$

And since the youngest child took 31 mangoes,

$$\frac{31}{420}R = 31$$

Hence,

$$R = 420$$

And since $R = \frac{5}{6}T$,

$$T = \frac{6}{5}R = \frac{6}{5}(420) = 504$$

Using practice questions such as this, the coach upholds the task of raising the level of the teachers' mathematics interest and competence.

A good coach can introduce topics on the history of mathematics such as the following: Arab writers dominated the scene of algebraists in ancient times for many centuries. The greatest Arab writer on algebra was Mohammed ibn Musa al-Khwarizmi (ca. 825), from whose book *Al-Jabr W'al Mukabala* the name *algebra* is derived. His major contribution was the introduction of the use of algorithms as a mathematical tool. Incidentally, the word *algorithm* is derived from the name *al-Khwarizmi.*

Al-Khwarizmi's work became known in Europe in the Twelfth Century. By the Sixteenth Century, the symbols we now use in algebra had already been, for the most part, slowly and painstakingly developed. Then there are little things that can be of interest to students, and such things a math coach should be prepared to share with the teachers, such as the "=" sign was first introduced by Robert Recorde in his 1557 book *The Whetstone of Witte.* In 1637, René Descartes introduced the exponent notation that we seem to take for granted today.

A math coach might also inform teachers of historical aspects of mathematics. For example, quadratic equations and their solutions played a special role in the history of algebra. The Babylonians first solved certain forms of quadratic equations 3600 years ago, as found in tablets that are now in a collection, YBC 6967, at Yale University. A translation of the problem on these tablets is equivalent to "Find the dimensions of a rectangle whose length exceeds its width by 7 and whose area is 60." The Babylonian solution uses no algebraic notation but serves as a prototype for similar problems. The solution to the Babylonian quadratic can be found in *Teaching Secondary Mathematics: Techniques and Enrichment Units*, by A. S. Posamentier and B. S. Smith, 9th Edition (Pearson, 2015). This book also provides a plethora (125 units) of other enrichment ideas which you as coach might want to embrace so that you can select the appropriate ideas for the teachers you are coaching.

5.1. Historical Notes

The embellishment resulting from the inclusion of historical notes in a mathematics lesson may be considered enrichment. Students usually get a more complete appreciation of a topic if they can relate to its origin. Any book on the history of mathematics can provide useful ideas for enriching your instruction by the inclusion of short anecdotes from the history of mathematics.

For example, one may ask: What do Pythagoras, Euclid, and James A. Garfield, the 20th president of the United States, have in common? The answer is: They each produced a proof of the Pythagorean Theorem. Currently, there are over 400 proofs of the Pythagorean theorem which a teacher can use — either from history or from current developments — to enhance the presentation of the most famous theorem. Some demonstrations of the Pythagorean Theorem have been done by people not known for their mathematical prowess, such as President Garfield, who in 1876, while still a member of the House of Representatives, produced the following proof[1]. Garfield was previously a professor of classics and has the

[1]Garfield, James. A. "Pons Asinorum." *New England J. Educ.* **3**, 161, 1876.

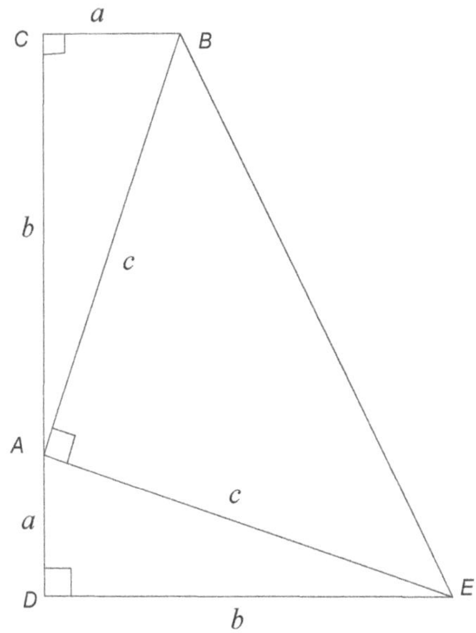

Figure 5.1

distinction today of being the only sitting member of the House of Representatives to have been elected president of the United States. Let's take a look at the proof he discovered.

In Figure 5.1, $\Delta ABC \cong \Delta EAD$, and all three triangles in the diagram are right triangles. Recall that the area of the trapezoid[2] *DCBE* is half the product of the altitude $(a + b)$ and the sum of the bases $(a + b)$, which we can write as $\frac{1}{2}(a+b)^2$.

We can also get the area of the trapezoid *DCBE* by finding the sum of the areas of each of the three right triangles:

$$\frac{1}{2}ab+\frac{1}{2}ab+\frac{1}{2}c^2$$

$$2\left(\frac{1}{2}ab\right)+\frac{1}{2}c^2$$

[2]A *trapezoid* is a quadrilateral with exactly one pair of opposite sides parallel.

We can then equate the two expressions that each represent the area of the entire trapezoid:

$$2\left(\frac{1}{2}ab\right)+\frac{1}{2}c^2=\frac{1}{2}(a+b)^2$$

This can be simplified to:

$$2ab+c^2=(a+b)^2$$

$$2ab+c^2=a^2+2ab+b^2$$

$$c^2=a^2+b^2$$

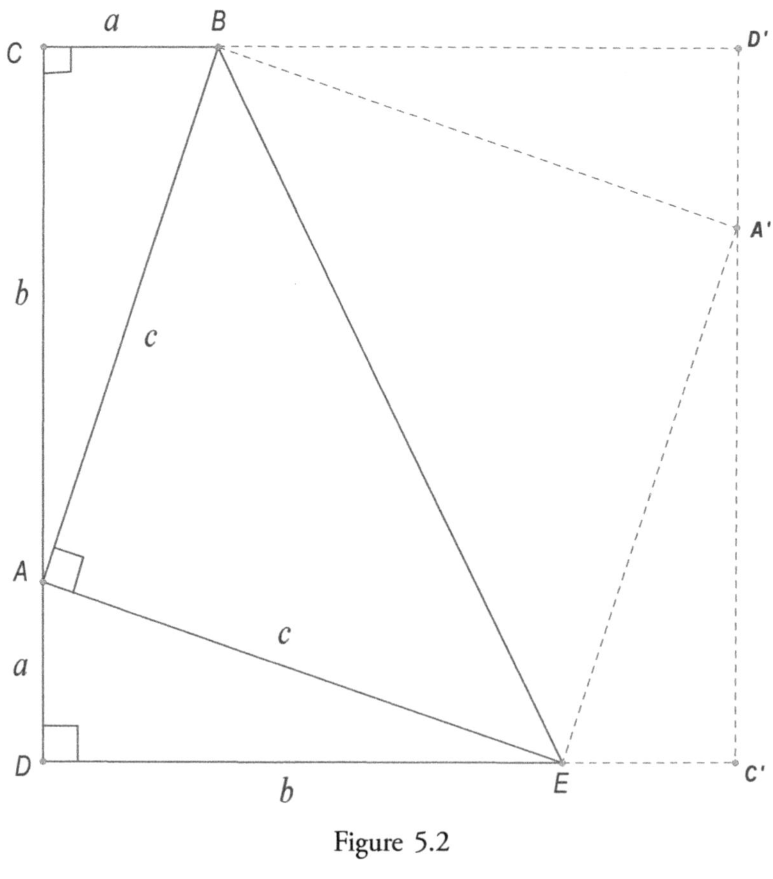

Figure 5.2

This is the Pythagorean Theorem as applied to right triangle *ABC*.

An informed reader may notice that Garfield's proof is somewhat similar to the one believed to be used by Pythagoras. If we "complete" a square from the given trapezoid in Figure 5.1, we get a "completed square", as shown in Figure 5.2.

A high school student[3] further manipulated this configuration and came up with yet another proof of the Pythagorean Theorem. She flipped[4] the figure of Garfield's proof (Figure 5.1) to get Figure 5.3.

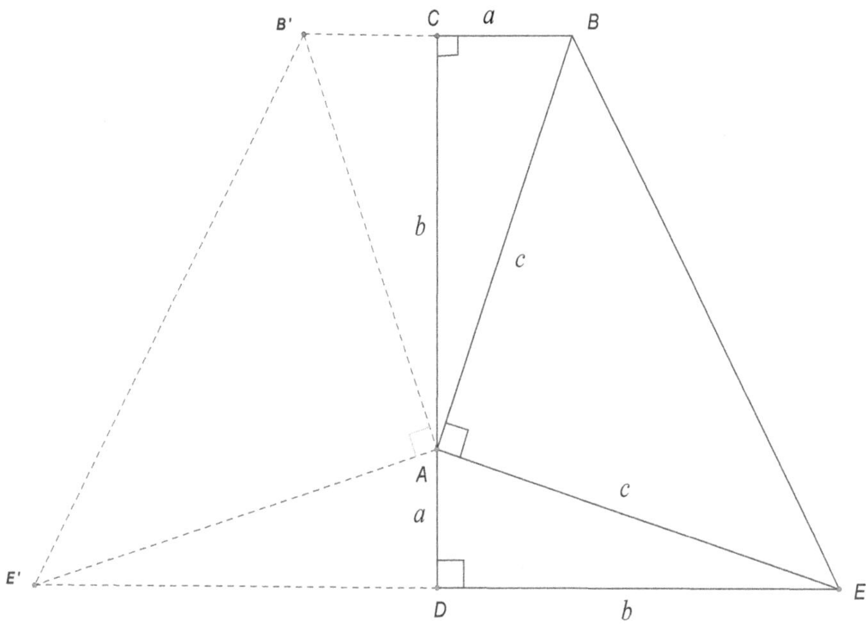

Figure 5.3

[3]Jamie deLemos, a student at Framingham High School in Framingham, MA presented her proof in *The Mathematics Teacher* (Vol. 88, No. 1, January 1995, p. 79).

[4]This is also called a reflection in line *CAD*.

She simply found the area of the trapezoid as the sum of the areas of the various right triangles. First, the area of the trapezoid $BB'E'E$ equal to one-half the product of its altitude and the sum of its bases, namely,

$$\frac{1}{2}(a+b)(2a+2b).$$

Now the area of this trapezoid is also the sum of the six right triangles that comprise it. That is, four times the area of triangle *ABC* and twice the area of triangle *ABE*, that is,

$$4\left(\frac{1}{2}ab\right)+2\left(\frac{1}{2}c^2\right).$$

Equating these two expressions for the same area (trapezoid $BB'E'E$), we get:

$$\frac{1}{2}(a+b)(2a+2b)=4\left(\frac{1}{2}ab\right)+2\left(\frac{1}{2}c^2\right).$$

Then simplifying, we get:

$$(a+b)^2 = 2ab+c^2$$

$$a^2+2ab+b^2 = 2ab+c^2$$

$$a^2+b^2 = c^2.$$

As the student herself says: "Who would have guessed that an average geometry student could come up with an original derivation of this famous theorem? Logic and creativity are the keys, plus a little push and some thinking." This might serve to motivate the reader to seek other proofs of the Pythagorean Theorem.

The following books are the sorts of references that a coach should be recommending to teachers to enhance their presentations.

1. Beckmann, Petr. *A History of π*. New York: St. Martin's Press, 1971.

2. Bell, E. T. *Mathematics, Queen and Servant of Science.* Washington, DC: Mathematical Association of America, 1979.
3. Bell, E. T. *Men of Mathematics.* New York: Simon & Schuster, 1937.
4. Boyer, Carl B. *A History of Mathematics.* New York: John Wiley, 1968.
5. Burton, David M. *The History of Mathematics, An Introduction.* Boston: Allyn and Bacon, 1985.
6. Campbell, Douglas M., and John C. Higgins, Eds. *Mathematics: People, Problems, Results.* Belmont, CA: Wadsworth, 1984.
7. Dunham, William. *Journey Through Genius.* New York: Wiley, 1990.
8. Eves, Howard W. *An Introduction to the History of Mathematics.* New York: Saunders College Publishing, 1983.
9. ________. *The Other Side of the Equation.* Boston: Prindle, Weber, and Schmidt, 1971.
10. ________. *In Mathematical Circles,* Vols. 1 and 2. Boston: Prindle, Weber, and Schmidt, 1969.
11. ________. *Mathematical Circles Revisited.* Boston: Prindle, Weber, and Schmidt, 1971.
12. Gamon, George. *One, Two, Three … Infinity.* New York: Viking Press, 1947.
13. Gray, Shirley B., and C. Edward Sandifer. "The Sumario Compendioso: The New World's First Mathematics Book." *Mathematics Teacher* 94 (2001): 98–103.
14. Kasner, E., and J. Newman. *Mathematics and the Imagination.* New York: Simon and Schuster, 1940.
15. Katz, Victor J. *A History of Mathematics: An Introduction.* Addison-Wesley, 2009.
16. Kelley, Loretta. "A Mathematical History Tour." *Mathematics Teacher* 93 (2000): 14–17.
17. Lee-Chua, Queena N. "Mathematics in Tribal Philippines and Other Societies in the South Pacific." *Mathematics Teacher* 94 (2001): 50–55.

18. *Mathematics Teacher* 98 (November 2000), the entire issue.
19. Moritz, R. E. *On Mathematics: A Collection of Witty, Profound, Amusing Passages about Mathematics and Mathematicians.* New York: Dover, 1958.
20. National Council of Teachers of Mathematics. *Historical Topics for the Mathematics Classroom.* 31st Yearbook. Reston, VA: NCTM, 1969.
21. Norwood, Rick. "A Star to Guide Us." *Mathematics Teacher* 92 (1999): 100–101.
22. Ostermann, A. and G. Wanner, *Geometry by Its History,* New York: Springer-Verlag, 2012.
23. Posamentier, Alfred S., and Noam Gordon. "An Astounding Revelation on the History of π" *Mathematics Teacher* 77 (1984): 52.
24. Posamentier, Alfred S., and I. Lehmann, π: A Biography of the World's Most Mysterious Number. Amherst: Prometheus Books, 2004.
25. Posamentier, Alfred S., and I. Lehmann, The Fabulous Fibonacci Numbers, Amherst: Prometheus Books, 2007.
26. Posamentier, Alfred S., and I. Lehmann, The Glorious Golden Ratio, Amherst: Prometheus Books, 2012
27. Posamentier, Alfred S. *The Pythagorean Theorem,* Amherst: Prometheus Books, 2010.
28. Resnikoff, H. L., and R. O. Wells, Jr. *Mathematics in Civilization.* New York: Dover, 1984.
29. Schaaf, William L., Ed. *Our Mathematical Heritage.* New York: Macmillan, 1963.
30. Stillwell, John. *Mathematics and Its History.* New York: Springer-Verlag, 1989.
31. Turnbull, H. W. *The Great Mathematicians.* New York: New York University Press, 1961.
32. Veljan, Darko. "The 2500-Year-Old Pythagorean Theorem." *Mathematics Magazine* 73 (2000): 259– 272.

33. Willerding, Margaret. *Mathematical Concepts, A Historical Approach.* Boston: Prindle, Weber, and Schmidt, 1967.

Some of these books will lend themselves better to quick extraction of anecdotes for classroom use; others will require a somewhat more thorough reading. Either way, these books should provide you and your fellow teachers with ample resources for enrichment through historical anecdotes.

5.2. Enrichment Techniques for All Levels

The core curriculum of most mathematics programs propose that all students be exposed to the same mathematics, though differentiated by levels of abstraction of tasks and concepts. Opportunities for enrichment of instruction exist at every level, particularly when integrated with real-life applications. Flexibility is the key to productive enrichment.

When enrichment activities are reserved for small groups, each particular group can learn at its own pace. In such groups, students will delve into ideas with classmates who share similar interests. They will have an opportunity to look more deeply into, and even beyond, those ideas developed during regular class sessions. Thus, they will make more challenging explorations and broader connections within and outside of mathematics. Teachers who are first beginning to teach may appreciate some support from the math coach in getting this organized. So be prepared!

5.3. Some Relevant Applications

Realistic and relevant applications of the mathematics being studied generally can provide an excellent source for enrichment. Here, we offer some support through the following examples.

Addition and multiplication of fractions can be enriched by computing simple probabilities after developing some class-made probability games (e.g., selecting cards from a deck, tossing a coin or die). The teacher's creativity is very important in making this a success. As coach, you may want to demonstrate how this could be done effectively.

Percentages can be exhibited through the computation and verification of newspaper advertisements involving percentages (e.g., sales, bank advertisements). Handled properly, these investigations can be both motivating and revealing, as some astonishing results often appear. A coach would be wise to have an arsenal of such material available to show teachers and students some of the more spectacular results.

Using formulas can have students compute the areas of real floor plans, or compute bank interest, or compute the volume of real but irregular shapes. When possible, these computations should take place in a realistic setting so as to take on more meaning for the students. Again, the coach should have some sample situations at hand.

Reading tables can be enriched by having students consider realistic problems using mileage charts, postal rate schedules, telephone rate schedules, bus schedules, income tax schedules, etc. Where possible, these should relate to local situations.

Basic concepts of geometry can be enhanced by having students do some direct and indirect measurement of local structures. Students should be asked to compute areas, volume, etc., from the data gathered.

Recreational Mathematics is another form of enrichment in mathematics. Anything "recreational" is generally of interest to students, especially if it is presented in the right tone. Therefore, mathematics topics being studied can be reinforced enthusiastically under the guise of recreation. It is important for the coach to impress upon teachers that the fashion in which they present anything that is recreational should be done in a manner different from the normal instructional methods.

Reducing fractions is often a topic that lacks any excitement. As coach you might want to introduce some divisibility rules for easier recognition of common factors, such as if the sum of the digits of the number is divisible by 3 or 9, then the actual number is divisible by 3 or 9.

Practice with addition which on the surface would appear to be a topic that lacks promise for any enrichment activity, might have you consider having a class work with magic squares of various kinds, as that can be used to reinforce addition facts through an unusual drill. This can lead to other forms of computation drill enrichment, by using various games and puzzles that require students to do computations as part of the activity can enrich an otherwise boring topic. These may be competitive games or puzzles that are individual activities. Here, the enrichment activity can be very instrumental in the class's ultimate success with the regular work, since a motivated student should perform better than one "turned off" to mathematics.

There are some activities that enable students to practice simple arithmetic and at the same time can be rather entertaining. As coach, you might introduce the following activity which will lead students to the same result regardless of which number they begin with.

Here is how this very unusual example begins. Follow the steps below:

1. Begin by selecting a four-digit number (except one that has all digits the same).
2. Rearrange the digits of the number so that they form the largest number possible. (That means write the number with the digits in descending order.)

3. Then rearrange the digits of the number so that they form the smallest number possible. (That means write the number with the digits in ascending order.)
4. Subtract these two numbers (obviously, the smaller from the larger).
5. Take this difference and continue the process, over and over and over, until you notice something disturbing happening. Don't give up before something unusual happens.

Eventually, you will arrive at the number **6174**, perhaps after one subtraction, or after several subtractions. When you do, you will find yourself in an endless loop. That is, if you continue this process with the number 6174, you will end up with that number again. When you have reached the loop, remember that you began with an arbitrarily selected number. Isn't this quite an astonishing result? Some teachers might be motivated to investigate this further. Others will just sit back in awe. Either way, they have been charmed again with the beauty of mathematics.

Here is an example of how this works with our arbitrarily selected starting number 3203.

The *largest* number formed with these digits is:	3320
The *smallest* number formed with these digits is:	0233
The difference is:	3087

Now using this number, 3087, we continue the process:

The largest number formed with these digits is:	8730
The smallest number formed with these digits is:	0378
The difference is:	8352

Again, we repeat the process.

The largest number formed with these digits is:	8532
The smallest number formed with these digits is:	2358
The difference is:	6174

The largest number formed with these digits is:	7641
The smallest number formed with these digits is:	1467
The difference is:	6174

And so, the loop is formed, since you keep on getting 6174. Remember, all this began with an arbitrarily selected number and will always end up with the number 6174, which then gets you into an endless loop (i.e. we continuously get back to 6174). In the practice there will be entertainment because some students will get to the 6174 sooner than others. It all depends on which number you selected at the outset.

One can also give students an insight into mathematics through the recreation of alphametics. Let's consider the following as an example of what a coach might keep in his arsenal of tools.

One of the great strides made by western civilization (which was adopted from the Arabic civilization) was the use of a place value system for our arithmetic. Working with Roman numerals was not only cumbersome but made many algorithms impossible. The first appearance of the Hindu-Arabic numerals, as mentioned earlier, was in Fibonacci's book, *Liber abaci*, in the year 1202. Beyond its usefulness, the place value system can also provide us with some recreational mathematics that can stretch our understanding and facility with the place value system and the algorithms we use with it.

Applying reasoning skills to analyzing an addition algorithm situation can be very important in training mathematical thinking. Consider the following, where the letters represent the digits of a simple addition:

$$\begin{array}{r} \mathbf{S\ E\ N\ D} \\ +\quad \mathbf{M\ O\ R\ E} \\ \hline =\mathbf{M\ O\ N\ E\ Y} \end{array}$$

Find the digits that are represented by the letters to make this addition correct. Show that the solution is unique.

Most important in this activity is the analysis, and particular attention should be given to the reasoning used. At first this may appear to be a daunting task, but if you take it step by step, you will find it rewarding and entertaining.

The sum of two four-digit numbers cannot yield a number greater than 19999. Therefore,

$$M = 1.$$

We then have

$$MORE < 2000$$

and

$$SEND < 10000.$$

It follows that

$$MONEY < 12000.$$

Thus, O can be either 0 or 1. But the 1 is already used; therefore,

$$O = 0.$$

We now have:

$$\begin{array}{r} \mathbf{S\ E\ N\ D} \\ +\quad \mathbf{1\ 0\ R\ E} \\ \hline =\ \mathbf{1\ 0\ N\ E\ Y} \end{array}$$

Now,

$$MORE < 1100.$$

If SEND < 9000, then

$$MONEY < 10100,$$

which would imply that

$$N = 0.$$

But this cannot be, since 0 was already used; therefore, it must be true that

$$SEND > 9000,$$

which implies

$$S = 9.$$

We now have:

$$\begin{array}{r} \mathbf{9\,E\,N\,D} \\ + \quad \mathbf{1\,0\,R\,E} \\ \hline = \mathbf{1\,0\,N\,E\,Y} \end{array}$$

The remaining digits from which we may complete the problem are

$$\{2,3,4,5,6,7,8\}.$$

Let us examine the digits in the units or ones column. The greatest sum is

$$7 + 8 = 15$$

and the least sum is

$$2 + 3 = 5.$$

If $D + E < 10$, then

$$D + E = Y$$

with no carry over into the tens column. Otherwise,

$$D + E = Y + 10,$$

with a 1 carried over to the tens column.

Taking this argument one step further to the tens column, we get

$$N + R = E,$$

with no carryover, or

$$N + R = E + 10,$$

with a carryover of 1 to the hundreds column. However, if there is no carryover to the hundreds column, then

$$E + 0 = N,$$

which implies erroneously that

$$E = N.$$

This is not permissible. Therefore, there must be a carryover to the hundreds column. So, $N + R = E + 10$, and $E + 0 + 1 = N$, or simplifying,

$$E + 1 = N.$$

Substituting this value for N into the previous equation we get:

$$(E + 1) + R = E + 10,$$

which implies that

$$R = 9.$$

But this has already been used for the value of S. We must try a different approach.

We shall assume, therefore, that $D + E = Y + 10$, since we apparently need a carryover into the tens column, where we just reached a dead end. Now the sum in the tens column is

$$1 + 2 + 3 < 1 + N + R < 1 + 7 + 8.$$

If, however, $1 + N + R < 10$, there will be no carryover to the hundreds column, leaving the previous dilemma of $E = N$, which is not allowed. We then have

$$1 + N + R = E + 10,$$

which insures the needed carryover to the hundreds column. Where $1 + E + 0 = N$, or $E + 1 = N$. Substituting this in the above equation $1 + N + R = E + 10$ gives us

$$1 + (E + 1) + R = E + 10,$$

or

$$R = 8.$$

We now have:

$$\begin{array}{r} \mathbf{9\,E\,N\,D} \\ +\quad \mathbf{1\,0\,8\,E} \\ \hline =\ \mathbf{1\,0\,N\,E\,Y} \end{array}$$

From the remaining list of available digits, we find that

$$D + E < 14.$$

So, from the equation $D + E = Y + 10$, Y is either 2 or 3. If $Y = 3$, then

$$D + E = 13,$$

implying that the digits D and E can take on only 6 or 7. If $D = 6$ and $E = 7$, then from the previous equation $E + 1 = N$, we would have

$$N = 8,$$

which is unacceptable since $R = 8$. If $D = 7$ and $E = 8$, then from the previous equation $E + 1 = N$, we would have

$$N = 9,$$

which is unacceptable since $S = 9$. Therefore,

$$Y = 2.$$

We now have:

$$\begin{array}{r} 9\,E\,N\,D \\ +\quad 1\,0\,8\,E \\ \hline =1\,0\,N\,E\,2 \end{array}$$

Thus

$$D + E = 12.$$

The only way to get this sum is with 5 and 7. If E = 7, we once again get from E + 1 = N, the contradictory N = 8, which is not acceptable. Therefore,

$$D = 7$$

and

$$E = 5.$$

We can now again use the equation E+ 1 = N to get

$$N = 6.$$

Finally, we get the solution:

$$\begin{array}{r} 9\,5\,6\,7 \\ +\quad 1\,0\,8\,5 \\ \hline =1\,0\,6\,5\,2 \end{array}$$

This rather strenuous activity should provide some important training in strengthening one's facility in mathematics, while being entertainment with students at the same time.

It is easy to fall into a trap of letting the recreational aspects of a lesson dominate the session. When taken to this extreme, the introduction of recreational topics can become counterproductive; any attempt to return to the intended curriculum will be viewed by the students with displeasure. Remember, the recreational aspect of a lesson is designed merely to enrich, not to replace, the regular curriculum.

5.4. Field Trips

As a math coach, one needs to think of alternatives to the normal classroom routines. It is uncommon to think of field trips as a way to enhance the mathematical instructional program. Yet such activities are not reserved only for the social studies class, for there are sites that can also enhance the mathematics taught. For example, if probability is the topic studied, a trip to a nearby racetrack, where betting is supposedly based on the likelihood of a horse winning, would be interesting to visit. Classes have been known to visit the backstage areas to see how the betting odds are calculated.

The more usual sites to visit would be to see how mathematics is used in industry would be in: actuarial work, engineering, architecture, business, computer programming, etc. In each case before a field trip is made, there should be a thorough preparation session for the class. The preparation session is sometimes almost as valuable as the trip. The teacher can spark students' interest by using a variety of audiovisual aids, including the Internet site of the place of visit.

As the coach for mathematics, you might want to emphasize the importance for teachers to preplan a field trip with the class. When students participate in the planning, they know precisely what the purpose of the trip is and what the expected outcomes are. Consequently, during the trip they will focus their time properly. Almost as important as planning is the post-visit session that reviews all that has been or should have been learned from the trip. Done properly, a field trip can enhance and make real the mathematics being taught.

5.5. The Internet

There are boundless applications on the Internet that will enrich an instructional program. A listing of websites, however, can quickly become outdated, for new sites appear daily and some older ones disappear. Local teacher organizations and the NCTM (https://www.nctm.org) are good sources for current sites. Relevant Internet sites can:

- provide historical notes
- share teaching ideas,
- offer exemplary lessons (enhanced by video clips),
- list ideas for enrichment activities,
- provide challenging problems,
- suggest connections with math students all over the world with whom to share ideas,
- provide video clips of applications of mathematics,
- suggest and guide gifted students through mathematics research activities,
- indicate connections between various areas and/or topics in mathematics, and
- many other ideas as boundless as your imagination!

5.6. The Gifted Student

Gifted students in mathematics are usually described as exhibiting ingenuity, intellectual curiosity, creative talent, an ability to assimilate and generalize, and a high level of mathematical achievement. A math coach must be prepared to also offer advice about alternatives for gifted students. Such students usually will participate in many extracurricular activities in mathematics. They are also more apt to read mathematics books, periodicals, and pamphlets. These independent activities may lead the more motivated among them to pursue a study of topics in mathematics that are either "off the beaten path" or are a part of some more advanced course (likely to be studied at a later date). Perhaps one of the more rewarding experiences for a teacher of gifted students is to observe a student making a discovery or developing an unusual approach to a topic or problem. This rare insight possessed by the gifted should be nurtured by the teacher through appropriately selected enrichment activities.

Let us once again review some enrichment activities, however this time more directed for gifted students. Once again, they can be categorized into three types: ***acceleration***, ***expansion***, and ***digression***. We will look at them individually as follows:

Acceleration usually involves moving gifted students along at a faster than usual pace. This might mean introducing elementary algebra earlier and then enabling the student to reach calculus (and sometimes beyond) while still in high school. It could also mean that full-year courses would be covered in less time, making room for a study of more advanced topics earlier. Perhaps the main advantage of this practice is that it allows the gifted student to remain appropriately challenged and thereby prevents loss of interest resulting from an otherwise lockstep curriculum geared to the average-ability student.

But there are also possible dangers to watch for when accelerating gifted students. If the acceleration is too rapid, the student may be asked to work with too much abstraction too quickly or too soon, that is, before he or she is actually ready for it. This sort of experience could be counterproductive and have long-lasting effects. The teacher must not only look for student readiness, but must also be heedful of the motivation that any student needs, regardless of ability. A student, after having been identified as gifted, should not merely be *pushed* along a predetermined track. If a proper balance between interest and ability is not achieved, there exists a likelihood that this gifted student could be turned off to mathematics.

Regardless of the form and degree of the acceleration, careful monitoring of the student both academically and socially is of paramount importance. The teacher must be sensitive to signs that may indicate that a student is perhaps mathematically overextended, generally overworked, or improperly placed socially. With this prescribed care, the acceleration process can be worthwhile; without such care it could be irreparably harmful.

Expansion refers to that form of enrichment which allows students to delve more deeply into the topics being studied. Such expansion of the regular mathematics curriculum would take place primarily as part of the classroom mathematics instruction, but it might be a part of an extracurricular program as well. As the math coach you might want to provide a few examples; such as when introducing the Pythagorean theorem at the upper grade levels, there are countless applications — over 400 proofs currently available of this theorem, many of which can be easily seen in the lower grades because they are largely visual. Another example of expansion would be to indicate the origin of some of the mathematical symbols and how they came to be universally accepted. There are many books that will help teachers come up with ideas, and as coach, it would be a good idea for you to collect an arsenal of such books at your ready reference, as we will recommend at the conclusion of this chapter.

In each of the preceding examples of enrichment through the expansion of a topic in the regular curriculum, you will notice that the discussion was always confined to an embellishment of the original topic and not a digression to another topic. This latter situation is our last form of enrichment for the gifted student to be discussed.

Our third form of enrichment referred to as a ***digression*** from the regular curriculum, where the teacher digresses from a topic to consider another topic that is related to, and generally an outgrowth of, the original topic. Since a gifted class or group of students can usually master a topic more quickly than an average class, more time is available for discussion of another related topic before continuing with the regular prescribed curriculum. These digressions are usually rewarding and can last from a portion of a class session to several sessions. As with other enrichment activities discussed, digressions should be used to enrich the regular curriculum, not detract from it. Often these enrichment activities can be glamorous, which by contrast can make the regular curriculum appear somewhat less appealing. Therefore, the teacher, conscious of this possibility, should try continually to relate a digression to the regular curriculum so that instead of detracting from it, the enrichment topic will in fact embellish it. The following examples should help provide a better understanding of what is meant by an enrichment activity that represents a digression from the regular curriculum. As math coach, it would be wise to have collected some ideas of digressions from the curriculum in your school.

For example, treated appropriately, the topic of networks can be a worthwhile digression from the study of informal geometry. With the possible goal of investigating the Königsberg-Bridge problem, the study of networks can be exciting and can lead to other related topological investigations

The only restriction on the topics that make appropriate digressions for enrichment is the teacher's judgment. Here the math coach's guidance can be very useful as well, as there are many possible topics from which to choose. The selected digression should be related to the topic in the curriculum that is to serve as the takeoff point and should be properly planned so as to have a clear beginning, a logical conclusion, and, above all, a specific purpose.

To the math coach, we would like to make aware a resource that contains 125 enrichment units: *Teaching Secondary Mathematics: Techniques and Enrichment Units,* Ninth Edition, (By A. S. Posamentier and B. S. Smith) Boston: Pearson, 2015.

5.7. Summary

As math coach, you should encourage teachers to constantly gather materials and ideas for enriching their teaching of mathematics. Regardless of the ability level of the students, appropriate enrichment activities can always be found. In some cases, these enrichment activities may be harder to secure than in others, yet the process of searching for an appropriate topic can, in and of itself, prove to be of great benefit as it will guide teachers to make a genuine effort to enrich his or her instruction. Very often, these motivating enrichment experiences go a long way toward developing a new appreciation for mathematics among the weaker or average students while being instrumental in encouraging further study in mathematics among the above average and gifted.

Chapter 6

The Various Roles of the Math Coach

As a math coach, you will assume a variety of roles and be faced with many different tasks. Some you will be familiar with, others you may not recognize. Let's take a brief look at a few of these roles for which the math coach might be responsible.

6.1. The Administration

Throughout this book, we have made many suggestions for involving teachers in the ongoing process of teaching mathematics in the best possible way. We have tried to emphasize that your role as the math coach is to help, not criticize nor evaluate. But what is your role when involving the principal or other administrators? This relationship can easily make or break the success of a coach's work.

First of all, remember that it is ultimately the responsibility of the principal or other administrators to improve instruction in all subjects — of course, including mathematics. The more he or she sees good mathematics instruction taking place, the better it is for the entire school. The administrator must understand that mathematics consists of a set of intertwined relationships and concepts instead of just a series of individual skills. The actual skills and concepts are learned over time, not necessarily in one or two sessions; certainly not immediately. Teaching mathematics is as such a long-term process. One of your major responsibilities is to keep the administration up to date on what is taking place in the classroom, in the greater community and governmental regulations.

Perhaps you might suggest a weekly or monthly meeting with the principal or a designee, where your coaching process will be discussed, and assistance will be rendered in further focusing your efforts. This would provide the regular forum for updating the principal on what is happening in the school's mathematics program. Consider also inviting the principal to come into a classroom and work with a group of youngsters as part of an actual lesson. After all, most principals were previously classroom teachers and would probably enjoy the experience; they might even learn some exciting and interesting mathematics from the students. Of course, be sure your teacher knows the principal is coming to visit the lesson, as one of the worst things a coach can do is to surprise the teacher with an unscheduled visit by an administrator.

Provide a copy of any articles (especially those that are research-based) that will help the principal understand the type of support the teachers are receiving. The principal/administrator may not be familiar with the teacher/learner roles as you change curriculum content and teaching techniques for any "new" curriculum concepts and skills. You are directly responsible for building the bridge to connect the administrator and the teacher. Remember, if you ever need to purchase some new materials for the math program, or need permission to attend a conference or meeting or want to visit other schools, the administrator is the one you eventually must go to. Therefore, keeping the administration well informed of what is going on is absolutely essential. A good administrator should take an active role in the teaching activities that go on in the mathematics classroom, and as such, the coach-administrator relationship is crucial for success in improving the mathematics program.

6.2. A Classroom Resource Center

Provide the math teachers with a mathematics resource center by finding a vacant classroom you can convert. If no such classroom is available, try to find some space in your office or a place to where the teachers can come to expand their instructional ideas and obtain suggestions. This resource center might contain sets of textbooks that are different from the texts the students are working with. Such unconventional texts may present alternative instructional strategies. Publishers are usually happy to gift some of their newest materials and textbooks to a school math coach — naturally, in the hope of selling their products to the school in the future. Also, include hands-on materials for the individual teachers to borrow and use with their students, as mathematics can easily come alive in a hands-on setting. A concerted effort should be made to attend local, regional, and state conventions to become acquainted with the newest products on the market. There, you may be able to learn about new ideas and obtain samples of new materials for the teacher resource center. The aim is to keep the teachers you work with up-to-date and as such. it would be favorable for you to invite various teachers to join you at conferences that you believe would provide some refreshing ideas for instruction. Rotating these conference invitations among your faculty would give everyone a chance to experience new developments, new curriculum, and new innovative materials. These new ideas presented at conferences are usually very valuable and helpful for your fellow teachers, since many of these ideas are presented by actual classroom teachers who talk about what has succeeded for them in their teaching. The teachers who attend can then lead a group discussion for their colleagues when they return to school. The resource center should thus, provide a comfortable place for teachers to meet, to sit and chat with you, discuss new ideas and successful lessons, and try new hands-on materials before attempting to use them with their students.

6.3. The Students

Some students describe their math class as boring, stressful, useless, and a waste of time, others exhibit signs of math anxiety and math avoidance, and others may just tune out any lesson that occurs in the mathematics classroom. Naturally, there are also many students who are enthusiastic about learning mathematics, and of course they need to be nurtured along as well. One of the most important parts of your job is to help the students learn. This may involve trying different approaches to teaching mathematics, or interesting them with some new and unusual topics. If the students are kept involved and interested in what is going on, they will respond positively to their teacher's lessons. The key here is to motivate students, a topic we have covered earlier.

One approach that a teacher can try, is to use some of the stronger students to assist other classmates. Many students like to teach and learn from one another. If the class is divided into small groups, the teacher can attend to those groups of students where he or she is most needed, while teaching goes on in the other groups. This provides an excellent opportunity to involve their students of varying abilities in the instructional process. The more students lead the teaching-learning activity, the less the teacher needs to do with these student-leaders and the more time the teacher can devote to those students who need your help. At the same time, the student leaders may discover some mathematics they had overlooked. It is well known that when students instruct or assist other students, they greatly strengthen their own knowledge and skill with mathematics, on top of developing their confidence and ability to deliver instruction.

6.4. Demonstration Lessons

Most teachers do not have the luxury of free time to observe someone teach a model lesson. Yet, the very act of seeing a lesson taught by an experienced teacher can offer a great deal of positive feelings towards mathematics teaching. You can teach a part of a lesson, perhaps an opener — that is, a motivating piece for the ensuing lesson — and then let the classroom teacher take over. For example, an activity called "What's Wrong?" makes an excellent student participation activity to start a class. Present a problem together with a solution. However, make at least three errors in the solution. These can be errors in arithmetic, errors in place value, a wrong operation, copying some numbers incorrectly, an erroneous assumption and so on. Let the class work in small groups and try to find and correct the errors and then solve the corrected problem. The math coach may refer to the plethora of ideas for motivating instruction offered in Chapter 4.

For younger students, you may need not have more than one deliberate error. For example, here is a $2^{nd}/3^{rd}$ grade problem.

Marianne brought home four large boxes from the store. She put two smaller boxes in each large box. Marianne reasoned she had 4 × 2 or 8 boxes in all. Was Marianne correct? What's wrong?

Another starter for a lesson requiring thinking is one called "Which One?" Divide the class into groups of 3–4 students. Give each group a list of four items that may or may not be obviously related. Ask the group to decide which one does not belong with the others and why. For example, one list might contain the following four items:

Triangle Square Trapezoid Trapeze

One group might decide that the one which does not belong is the "Trapeze" since it is the only one not a geometric figure. Another group might say the one which does not belong is the "Square", since the other three words all begin with the letters "Tr". Both answers are correct, but be sure the group explains **why** they selected the one they did.

Another one might be to give them the following numbers:

2 3 9 11

Ask which one does not belong with the others and why. One possible answer would be to choose the "2" as different — the only even number. Another possible choice to select is "11" — the only two-digit number. Some students might select "9" as the different one — it's the only non-prime (composite) number. All these answers are correct when accompanied by the proper explanation! All involve the students in a discussion. Yet many students will still ask, "Which one is right?" It is rare that the students find themselves involved with a problem that has more than one correct answer. The students are always looking for THE correct answer. Once again it is the "why" that is most important.

6.5. Workshops

A workshop is a meeting of people (in this case math teachers) where an exchange of ideas takes place. Any person can oversee a workshop; the key to a good workshop is a well-planned *exchange* of ideas, where every person is expected to participate and contribute. All ideas offered (correct or incorrect) should be considered and discussed as they arise. It is the job of the workshop leader to keep everyone on track. For the most part, many math workshops that take place are arranged by and run by the coach. However, this need not always be so — any one of the teachers can work with you and arrange to lead a workshop. Having one of the teachers in charge will help instill confidence in the teacher's work. Often, some teachers may feel the need for a workshop on a particular topic, and will ask you to arrange a workshop for everyone. Here is where your ability to act as a resource person comes into play.

The success of any workshop lies in the ability of the coach to secure some background materials, so that the session does not lag. As the resource person in the department, you are expected to have a basic knowledge of the topic under consideration, but not to dominate the discussion unless your input is needed.

When you set up the workshop, be certain that those expected to participate are informed in advance of the topic of interest. This will give the teachers time to think about and prepare their contribution. At the same time, it will help keep the workshop on track, as you would not want the workshop to run too long — it most probably will take place after the regular school day and teachers, having worked the whole day, will be tired. Like any good lesson plan, you should have an interesting opening idea to get the participant's attention — essentially a motivating device! Here is a worksheet that can be used to begin a workshop on operations with decimals. It works well if the participants work in pairs before arriving at a final answer. No calculators, please.

There are 4 mistakes on this worksheet. Find them and correct them.

(a) $0.5 \times 3 = 0.015$
(b) $0.5 \times 0.3 = 0.15$
(c) $5 \times 0.3 = 1.5$
(d) $0.05 \times 0.03 = 0.015$
(e) $0.15 \div 0.3 = 0.05$
(f) $0.3 + 0.5 = 0.8$

Some careful arithmetic reveals that (a), (d), and (e) are incorrect. But where is the fourth error? Aha! Did anyone consider the numeral "4" in the directions? That too, is incorrect. It should be a "3". This little bit of humor can gain the participant's interest and start the workshop on a relaxed tone.

6.6. Brainstorming

A special kind of workshop is the brainstorming session, which encourages everyone to take an active role in the workshop. The only role decided in advance is that of the coach who should be in charge and acts as leader. Ideally, you should notify all the teachers of the topic to be considered so they can do some research (if they wish) and gain some insights in advance. However, in a brainstorming type of workshop, prior research is not necessary or even encouraged.

The session is designed to produce spontaneous ideas and suggestions to solve a problem. The atmosphere should be pressure-free, relaxed, and informal. Some of the ideas put forth by the participants may seem weird or very odd, and that's fine. All contributions and suggestions should be accepted and noted. There is no criticizing of ideas as "bad", "weird", "silly", or "dumb". There should also be no praise of what some consider a "superb" or "good" idea either. Just write all contributions down for later discussion. Encourage the teachers to build on ideas put forth by others. What begins as an odd idea might lead to a great suggestion. When the number of suggested ideas being offered starts to dwindle, the group should begin to consider all the items that were suggested to that point. Would a specific idea work? Would it go towards resolving the problem at hand? Does it lead to a new idea? Gradually, a refined list of positive suggestions should emerge as ideas are discussed, reworked and rewritten. Everyone has a part in the final decisions.

6.7. Outside Ideas

As the coach, you should always be on the lookout for new and different ideas to transmit to your fellow math teachers. But where does one find these ideas? Be sure you join your professional organizations such as the National Council of Teachers of Mathematics, and your local and state organizations. Examine their publications carefully, and look at the meeting programs for workshops being offered. As we have said before, try to attend some of those that seem most interesting and of value. You might even see and hear an exciting speaker, whom you can invite to do a similar session with the teachers at your school. A new voice and some new ideas are always stimulating. If you can persuade one or two teachers to actually attend a professional conference, it will generate a great deal of excitement. Those who actually attend the meeting should bring back a great deal of new material to share. Even visiting another school and talking with their math coach can provide you with some new material to pass along to your team.

6.8. The Teachers

The most readily available resources to the coach are the teachers being coached. All of your fellow teachers are professionals, and all bring some important mathematics teaching to their colleagues. The teacher with 20 years of experience can lead a discussion on what has always worked in a particular lesson; the teacher fresh out of college can offer some of the latest techniques of teaching that a more experienced teacher may not have been aware of or may have forgotten. Typically, the younger teachers may be more technology savvy. One of your responsibilities is to leverage on the various backgrounds of your colleagues. If you have a teacher who enjoys teaching geometry, have that teacher prepare a brief talk on why teaching geometry is enjoyable. What is there that makes the study of geometry so interesting? What approach works with the students? A teacher who likes to make use of the latest technology, can do part of a workshop on how to use this technology to prepare a more effective instructional program. Neither of these teachers needs to prepare a full workshop. Rather, you can arrange a workshop potpourri of various items that are of importance to outstanding teaching. One of your jobs is to unearth this hidden talent among the staff and use it to the best advantage, such as putting together the above-mentioned meaningful workshop. Or, you might provide an opportunity for a few of the teachers to work together to discuss important topics in the curriculum — yet again, with you as guide.

For teachers who cannot attend, have some prepared notes on what occurred, as well as who conducted the workshop. Encourage the sharing of promising ideas, for teachers themselves are an excellent resource. Remember, teachers can modify a successful lesson to fit their individual style of teaching. Even if teachers take away one single idea, the workshop was a success!

Show interest in your colleagues' work. Once in a while, ask if they would mind your attending one of their classes that they themselves choose. Before the lesson starts, plan with the teacher even if it's only an informal chat. Help the teacher plan for some good questions to move the lesson along and help support the students' learning. After the lesson meet promptly with the teacher to discuss what went on. *Do not criticize*. Remember, there is no "best way" to teach the lesson — only "another" way.

When the teachers, who do not understand the mathematics they are attempting to teach, they tend to equate "teaching" with "covering ground"; merely meeting the written objectives of a lesson without being sure the students have learned and understand the material, is a wasteful use of time. If a student asks, "I know how to do it-just tell me if I multiply or divide?" when confronted with a word problem, one can be sure that this is a typical case of meaningless mastery of skills with no understanding of what they mean or when to use them.

Watch how much the students talk to others in the class during the lesson. We (and they) learn more when explaining what they were doing and why they did it. As we have said many times before, student-to-student discourse should be encouraged.

6.9. Testing

Another topic to be discussed with your fellow teachers is the role of testing. What can we learn from tests? Why do we test? First of all, there are two kinds of tests to consider.

- Standardized
- Teacher made

Teachers should know all about the standardized testing that takes place in their school and/or school district. These standardized tests play a significant role in your work. As mentioned previously, teachers should not just "teach to the test." Rather, have them consider the standardized tests as an opportunity to show how much their students have achieved. These tests offer an opportunity to see how their students compare with other students of the same age and ability. The tests can also reveal where specific weaknesses occur and help plan for future lessons. Unfortunately, many teachers feel that their job is in jeopardy unless their class does well in these tests. This is understandable, but not always true. A standardized test is developed by a test construction agency or group of selected coaches and teachers. It is then administered to a sample group of students and the results are carefully chosen or rejected. This "standardizes" the test so it should give the same kind of results when administered to the larger group of students throughout the district or state.

The teacher-made test is another aspect of teaching. Impress upon your fellow teachers that they should be very careful in constructing a test, for the teacher-designed test is not used to compare students with one another. Rather, it is to evaluate how much the individual student has achieved in learning concepts and skills, and some teachers use their teacher made tests as a self-assessment of their instructional success. Teachers should always begin constructing the test by determining which skills and concepts are to be tested and at what level of mastery. Above all, do not test the same concept or skill excessively on the same test. Emphasize that when constructing a test, it is not fair to penalize a student for getting this same concept wrong multiple times, if he or she can't get the concept correct once or twice.

Point out that multiple-choice questions are easy to grade, but rarely give them any information as to why the student got an answer correct or incorrect. For example, here is a multiple-choice question taken from a teacher-constructed test:

Joanna and Mark collected wild blueberries. Mark collected 50 berries. This was twice as many blueberries as Joanna collected. How many blueberries did they both collect altogether?

(a) 12.5 (b) 25 (c) 50 (d) 75 (e) 100

The answer to this question, either a, b, c, d, or e, can easily be graded as correct or incorrect. But, how did the students arrive at their choice? Was it by luck? (Since they have a 20% choice of just guessing correctly.) Was there an arithmetic error? Perhaps there was an error in concept. The simple algebraic thought processes might be missed by a child. Reiterate that the work the child does is just as important as the correct answer, if not even more so. A workshop on constructing a good test might be in order especially for the less-experienced teachers. Some have the attitude that "I'll stump them with this clever question." instead of "This will let them show me if they have grasped the concept." Even some of your more-experienced teachers can use a refresher workshop, as well as being shown some alternative test styles. Remember that the change in the goals of the current mathematics curriculum has resulted in requiring more carefully-constructed tests!

6.10. Fostering Creativity

Throughout this book, we have been referring to what is called "creativity" or "being creative". As a math coach, one of your primary duties is to help math teachers become more creative in their own teaching and, in turn, foster creative thinking in their students. Creative thinking is a major goal in today's mathematics classroom. Creativity or being creative refers to the ability to transcend traditional ideas, recognize patterns, and develop meaningful new ideas, new interpretations of given information and new methods of presenting this new information to others. But how do you *teach* creativity? Let's take a look at some ways of doing this.

6.10.1. *Record and reward unusual ideas*

Some teachers tend to discourage any responses they get from their students that lead away from their more traditional lessons. For example, when presenting a lesson on pattern discovery, the teacher asked for the next four terms of the sequence: **1, 2, 3, __, __, __, __, ...** Here are some of the responses given by students:

1, 2, 3, 1, 2, 3, 1, ...	(Repeating sequence)
1, 2, 3, 5, 8, 13, 21, ...	(Fibonacci-type sequence)
1, 2, 3, 3, 2, 1, 1, ...	(Forwards and backwards)
1, 2, 3, 6, 12, 24, 48, ...	(Sum of all previous terms)
1, 2, 3, 4, 5, 6, 7, ...	(Counting numbers)

The class then had a discussion (a bit heated as students tend to defend their choice) as to which one was correct. They finally agreed that *all* were correct depending on the pattern rule they were using to determine the terms of the sequence. Many agreed they would never have discovered some of these sequences by themselves. What started as a seemingly mundane task turned out to be a lesson involving creative thinking and expression. As a further exposure to what might be considered creative thinking, have them consider the following sequence: **1, 2, 4, 8, 16, ...** Clearly most people would guess that 32 is the next number. Yes, that would be fine in most cases. However, when the next number is given as **31** (instead of the expected 32), cries of "wrong!" are usually heard. Much to your amazement this is also a correct answer, and **1, 2, 4, 8, 16, 31** can be a legitimate sequence.

The task now is to convince students of the legitimacy of this sequence. It would be nice if it could be done geometrically as that would give convincing evidence of a physical nature. We will do that later, but in the meantime, let us first find the succeeding numbers in this "curious" sequence.

We shall set up a table of differences (i.e. a chart showing the differences between terms of a sequence), beginning with the given sequence up to 31, and then work backwards once a pattern is established (here at the third difference).

Original Sequence	**1**		**2**		**4**		**8**		**16**		**31**
First Difference		1		2		4		8		15	
Second Difference			1		2		4		7		
Third Difference				1		2		3			
Fourth Difference					**1**		**1**				

With the fourth differences forming a sequence of constants, we can reverse the process (turn the table upside down), and extend the third differences a few more steps with 4 and 5.

Fourth Difference					1		1		1		1				
Third Difference				1		2		3		**4**		**5**			
Second Difference			1		2		4		7		**11**		**16**		
First Difference		1		2		4		8		**15**		**26**		**42**	
Original Sequence	1		2		4		8		16		**31**		**57**		**99**

The bold-type numbers are those that were obtained by working backwards from the third difference sequence. Thus, the next numbers of the given sequence are: 57 and 99. The general term is a fourth-power expression since we had to go to the third differences to get a constant. The general n-th term is: $\dfrac{n^4 - 6n^3 + 23n^2 - 18n + 24}{24}$.

One should not think that this sequence is independent of other topics in mathematics.

Consider the Pascal triangle* shown in Figure 6.1.

1

1 1

1 2 1

1 3 3 1

1 4 6 4 1

1 5 10 10 5 1

1 6 15 20 15 6 1

1 7 21 35 35 21 7 1

1 8 28 56 70 56 28 8 1

Figure 6.1

Consider the horizontal sums of the rows of the Pascal triangle to the right of the bold line drawn, as shown in Figure 6.2 (overleaf) — our newly developed sequence is once again: **1, 2, 4, 8, 16, 31, 57, 99, 163, ….**

*This triangle is formed by beginning on top with 1, then the second row has 1, 1, then the third row is obtained by placing 1's at the end and adding the two numbers in the second row (1 + 1 = 2) to get the 2. The fourth row is obtained the same way. After the end 1's are placed, the 3's are gotten from the sum of the two numbers above (to the right and left), that is 1 + 2 = 3, and 2 + 1 = 3.

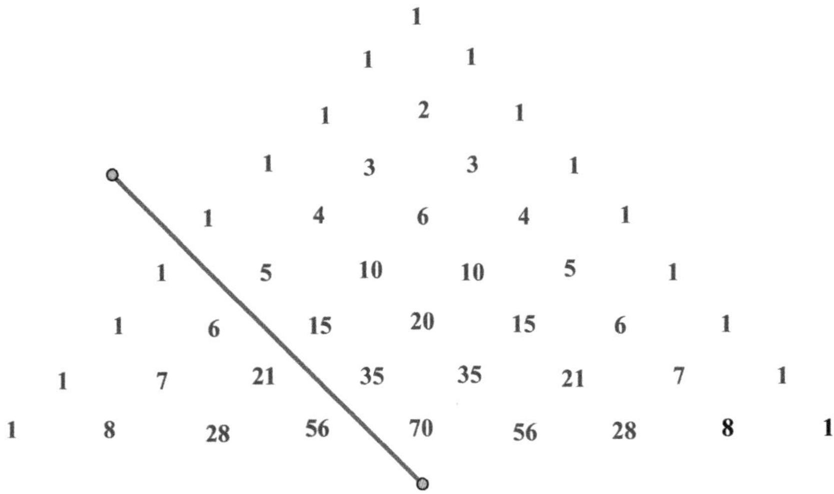

Figure 6.2

An alternative geometric interpretation should help convince your students and teachers of the beauty and consistency inherent in mathematics. To do this, we shall make a chart of the number of regions into which a circle can be partitioned by joining points on the circle. You ought to try this by partitioning a circle. Just make sure that no three lines meet at one point, or else you will lose a region.

Number of points on the circle	**Number of regions into which the circle is partitioned**
1	1
2	2
3	4
4	8
5	16
6	31
7	57
8	99

Now that you see that this unusual sequence appears in various other fields, you may feel a degree of satisfaction may be emerging. This is the sort of thing that can get students excited about math as they see that some expectations can lead to other more extensive connections within mathematics.

6.10.2. *Build a classroom atmosphere that encourages freedom of expression*

As long as the students can explain and justify why they answer as they do, their answer should be accepted and considered as correct. In the lesson described earlier involving the number sequence beginning with **1, 2, 3, ...** above, it is apparent that the students felt encouraged to express their answers without fear of being told "No! That's wrong!" The teachers should readily accept any answer that is correctly explained.

6.10.3. *Do not discourage controversy*

Students who discover one particular answer over others are often prone to argue quite vehemently over why their choice of answer is the correct one. Others may disagree and will defend their choice with equal vim and vigor. Teachers should let the argument continue for a while, until the discussion begins to die down or get out of hand. Above all, do not state which is correct, for they are all correct if the explanation is correct. A great deal of good mathematics can be obtained from such a controversial discussion, and this is a message that needs to be transmitted to all math teachers.

6.10.4. *Encourage cooperative learning when solving a problem*

Throughout this book, we have suggested to math coaches that their teachers encourage cooperative learning in small groups, which might involve having students learn from each other as they bounce ideas back and forth in an effort to solve a particular problem. Teachers should inform each group of students that the teacher will select one representative at random from each group to explain the group's solution or results. The group should finally agree on one solution that every group member can accept and explain.

6.10.5. *Encourage students to create new ideas, different interpretations, and new methods of arriving at a satisfactory solution*

While this suggestion may sound like it will create total chaos within the classroom, be assured it will not. Rather, it usually leads to exploration of ideas from a variety of viewpoints. It also helps transcend solving a problem in one traditional way. Students will have to learn to respect each other's ideas, thoughts, and contributions. It has been said by many educators that solving one problem five different ways is more productive than solving five problems each in one way.

Although there are lots of examples of problems that can be solved in a variety of ways, let's just take a very simple one that you can show without expecting the students to reach back for many other mathematical concepts. We shall consider the following problem along with some student solutions.

As shown in Figure 6.3, point E is selected anywhere on side CD of rectangle ABCD. Point E is then connected to points A and B to form a triangle AEB. The dimensions of rectangle ABCD are $AB = 12$ *and* $AD = 8$.

What is the sum of the areas of triangle AED and triangle BCE?

The most obvious solution is to recognize that both triangles are right triangles and the areas can be found using the formula that the area is one-half of the product of its legs.

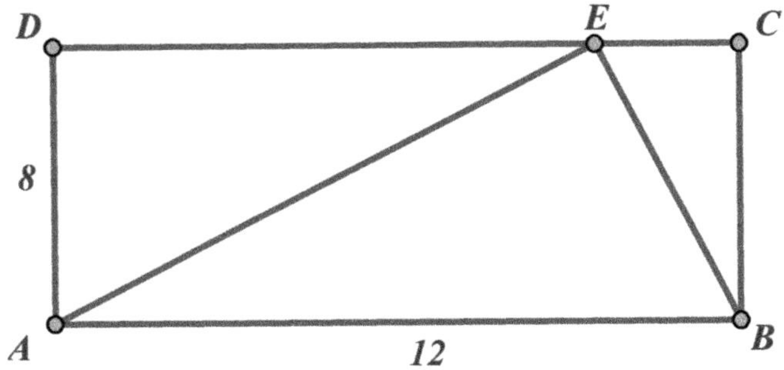

Figure 6.3

$$\text{Triangle } ADE = \frac{1}{2}(8)(DE) = 4\ DE,$$

$$\text{Triangle } BCE = \frac{1}{2}(8)(CE) = 4\ CE.$$

The sum of the two areas is

$4\ DE + 4\ CE = 4(DE + CE) = 4(CD) = 4(12) = 48$ square units.

Another student suggested dropping a perpendicular to AB from E, as shown in Figure 6.4 (overleaf). The area of the shaded triangle $ABE = \frac{1}{2}(AB)(8) = 48$ square units, which is one-half of the area of the rectangle whose area is $8 \times 12 = 96$. Therefore, the area of the two triangles is $96 - 48 = 48$.

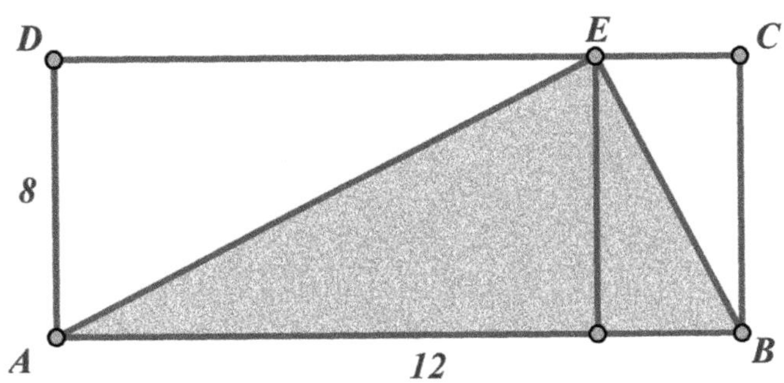

Figure 6.4

Another suggested solution asked to place E at the midpoint of CD, as shown in Figure 6.5. The areas of the two non-shaded right triangles are equal in area and each has an area of $\frac{1}{2}(8)(6) = 24$ square units. So, the combined non-shaded areas = 48.

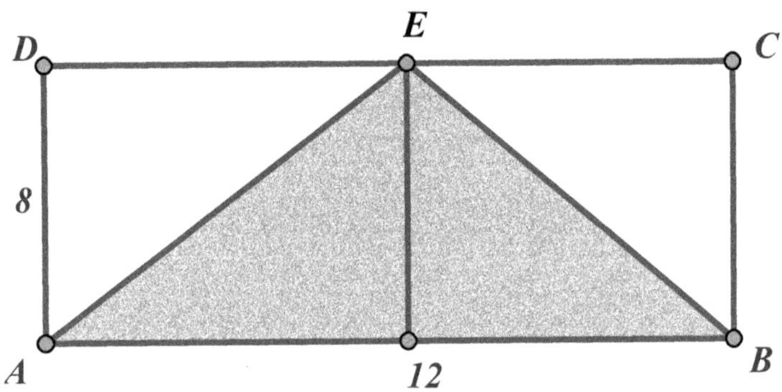

Figure 6.5

For our fourth solution procedure, we take point E, which was chosen at random and allowed one student to place point E at point C so they coincide, as shown in Figure 6.6 (overleaf). Thus, AE is a diagonal and divides the rectangle in half.

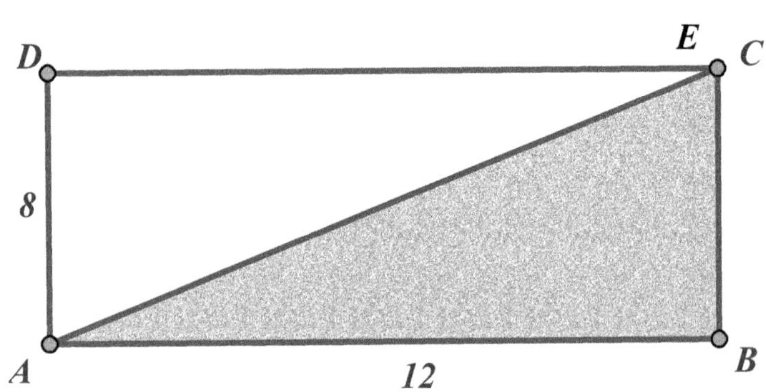

Figure 6.6

The area of each triangle is half the area of the rectangle, which is $\frac{1}{2}(8)(12)=48$ square units. The original two triangles are now represented by the one non-shaded triangle whose area is 48. Notice how each solution offers the opportunity to discuss some good mathematics.

Thus, the role of the math coach is to work with the math teachers to continuously stimulate them to try new ideas with the ultimate goal of making their instruction highly interesting and motivational. A stimulating teaching plan could involve introducing enrichment topics, which are not necessarily part of the standard curriculum, but stress problem-solving strategies with appropriately chosen applications from the real-world experiences, and actively involving the students throughout each and every class session.

Chapter 7

Some Final Words

7.1. General Instruction

Let's begin these final words with a caveat. The term "small groups" as we have been using it, takes on various meanings for various classes in various schools, and according to the teacher's skills. Teachers must understand how to divide their classes into different-sized groups to obtain the best results; no one kind of grouping will fit all situations. The topics and concepts being taught can easily require different groupings. Remember, not all teachers are used to teaching in the same way. Some still prefer the class as a whole group arrangement, while others are more comfortable making extensive use of various size groupings, with the teacher acting as a resource person.

In this age where the inclusion of all students is stressed, teachers must be certain that students of all abilities are involved in the learning process. Hence, some teachers may opt to group by ability, permitting a group of faster-learning students to work on advanced material by themselves with further assistance and enrichment by the teacher, while others might make a group of mixed abilities so that students can learn from each other. The tasks at hand may call for different kinds of groupings.

One of your most important tasks as a coach is to build a good relationship with the teachers you advise. Try to begin any meeting by mentioning some positive comments about a lesson that was observed. You might add some other ideas and perhaps suggest that the teacher try something different just to see if it succeeds. It would also be advantageous to mention that this would help you in your work to find out if an idea works well or not. You might even ask teachers if you can reverse roles; that is, if you can try teaching a lesson with their students and have the teachers critique you! Accept their critique readily and you might even learn some new ideas. At the same time, you are indicating that improving the educational practice is a team effort. This enables you to introduce a new performance idea without criticizing any teacher's lesson.

At faculty meetings, ask them to examine teaching strategies that have worked for each of them in the past, and have them identify strengths and weaknesses of each other's lesson and their techniques.

- Why did they work?
- Did the teacher arouse curiosity in their students?
- Did they challenge their students?
- If the strategy did not work, why didn't it?
- Did the teacher give the students sufficient time to think before reacting?
- Did they allow students to work with a partner in pairs? In small groups?
- Did the teacher pause when asking questions?
- Did they allow students to ask questions?
- Did they allow other students to try to answer each other's questions?
- When teaching an entire class as one group, did they at least move around the room and speak from various locations? Doing so allows students sitting in the back or on the sides of the room feel as though they are at the front of the room.

Build on positive techniques so all can discuss trying them in their own class.

Some teachers may be nervous when it comes to teaching mathematics (or just teaching in general), so try to strengthen their confidence and resolve towards teaching. Keep praising good lessons, good ideas, good techniques that work. But a word of warning: don't praise a lesson just to sing praise. Teachers can readily see through "phony" praise and start to ignore any praise whatsoever. Well-earned praise however, can boost their morale.

While trying to "popularize" mathematics, occasionally introduce topics and ideas that are not part of the curriculum and that can be easily used to motivate students. There are lots of sources for such ideas, and one such is "The Joy of Mathematics" (Prometheus Books, 2017); moreover, this publisher has a host of such books to offer some useful ideas to enhance and enrich instruction.

Above all, encourage teachers to develop a classroom atmosphere that is relaxed by perhaps decorating their classroom with mathematically appropriate posters, or even place the desks and chairs in different arrangements periodically. A colorful and cheerful classroom encourages student-to-student interaction — it involves the students taking an active role in discussion and does not become a teacher dominated lesson. We know that very often students learn more when working with one another rather than working directly with the teacher. You might encourage teachers to add some levity to their classes by having them do a short study of some of the more famous mathematicians from the past. One way of deciding how that might be is to list some of the birthdates of these mathematicians so that that could be a guide as to when and about whom such reports could be made. Here is a list of some mathematicians who might be included:

- Sonya Kovalevsky January 15, 1850
- Daniel Bernoulli February 8, 1700
- Emily Noether March 23, 1881
- Rene Descartes March 31, 1596
- Leonhard Euler April 15, 1707

- Carl Friedrich Gauss April 30, 1777
- Blaise Pascal June 19, 1623
- Bernhard Riemann September 17, 1826
- George Boole November 2, 1815
- Nicolai Lobachevsky December 1, 1792
- Isaac Newton December 25, 1642

Try to set up teaching events, possibly a workshop where you occasionally invite parents to sit in on a lesson so that the parents can feel a part of the class. It will also provide the parents with a better understanding of what their children are doing in class. Much of today's mathematics instruction is quite different from the experiences parents had when they went to school. When parents take part in the lesson, they become an additional resource for your students (and for you) to draw upon.

Start initiatives to make the math relevant to the teachers so that they, in turn, can make it more relevant to their students. This should actually be a team effort where you, as coach, would encourage the teachers to seek out real-world applications for the mathematics that they are currently teaching, and then share them with colleagues so that all can benefit from these findings. Oftentimes, an advertisement involving the use of mathematics provides an opportunity to gain student interest and greater insight into the situation which sometimes could be considered counterintuitive.

For example, two advertisements in a local newspaper recently raised an interesting issue for discussion. Is the interest earned on a 1-year Certificate of Deposit with an interest rate of 1.5% a year that requires a minimum deposit of $2500 better than the money earned at an interest rate of 1.25% but requires only a minimum deposit of $1500 and lasts for 18 months? This even raises the question of family finances, and might not be appropriate for everyone.

Another example could be where a store offers a 30% discount on a particular holiday in order to overshadow his competitor who constantly offers 20% discounts. When the second store sees this change, he offers an additional 10% discount on top of his already 20% discount. Which store gives you the best price? When numbers appear in newspapers and magazines, they offer opportunities for a mathematical investigation, which could be enlightening as well as instructive. Just emphasize the habit of looking for them.

7.2. Leveraging on Technology

In today's world, technology plays a major role in the lives of students. Encourage the use of technology in class, but stress that it should not be at the expense of the mathematics being taught. It is no secret that technology changes continuously, whereas curriculum changes periodically. Some teachers tend to use technology to guide their instructional program, while other teachers, seeing the curriculum foremost, instead search technological advances to see how the curriculum can best be supported. It is the latter that we believe is the better way to encourage teachers.

For example, there are a number of geometric supports for the computer. Construction programs such as Geometers' Sketchpad and Geogebra are incredibly useful in generating a true understanding of geometry. If you draw a random quadrilateral joining the midpoints of the sides consecutively, the resulting figure will always be a parallelogram and that can be shown by changing the shape of the original quadrilateral and noting that the end result will always be a parallelogram. Proper manipulation will also allow students to understand under which circumstances a parallelogram would become a square, a rectangle or a rhombus.

However, on the downside, one may argue that there is no advantage to using technology, if one might be better off just using the chalkboard. In other words, explain to teachers not to let students use the technology as a play tool when we would like it to be a support to instruction. Some teachers believe that if they can transmit ideas and concepts as easily on a chalkboard as they can on a SmartBoard, then perhaps the chalkboard would be better because fewer things can go wrong technologically. On the other hand, there are many advantages to using the SmartBoard, such as getting printouts of what was put on the screen or transmitting what was on the screen to student's computers. The coach has a responsibility of helping teachers sort out best advantages to using technology in the instructional program.

7.3. Applying History

Involving children in the history of mathematics makes the subject come alive — it adds a note of realism to the work they are doing in class. There are lots of anecdotes that can bring lessons to life and as coach it would be desirable for you to encourage teachers to search out historical topics that can be used to motivate a lesson or enhance a lesson through its historical application.

A very clever introduction to the sum of an arithmetic sequence is to tell the story of young Carl Friedrich Gauss, as seen on Page 26 in Chapter 2. Stories such as this would not only prove highly motivating to the class, but would also give some insight into the various beautiful relationships that mathematics offers. As coach, leading teachers to seek out relevant stories will certainly enhance instruction.

7.3.1. *What the Bible has as the value of π*

This particular story could prove to be quite enlightening and motivating for students and teachers alike. In the past, it was believed that the value of π in the Bible was equal to 3. But students may want to get a look at history from another point of view. For now, let's stay with this "crude" approximation of 3 for π. You'll be surprised to know that for centuries scholars believed that this was the value that π was to have had in biblical times. For many years, virtually all the books on the history of mathematics stated that in its earliest manifestation in history, namely in the Old Testament of the Bible, the value of π is given as 3. Yet recent "detective work" shows otherwise.[1]

One always relishes the notion that a hidden code can reveal long lost secrets. Such is the case with the common interpretation of the value of π in the Bible. There are two places in the Bible where the same sentence appears, identical in every way except for one word, which is spelled differently in the two citations. The description of a pool, or fountain, in King Solomon's temple is referred to in the passages found in 1 *Kings* 7:23 and 2 *Chronicles* 4:2, and reads as follows:

> **"And he made the molten sea[2] of ten cubits from brim to brim, round in compass, and the height thereof was five cubits; and *a line* of thirty cubits did compass it round about."**

The circular structure described here is said to have a circumference of 30 cubits[3] and a diameter of 10 cubits. From this we notice that the Bible has $\pi = \frac{30}{10} = 3$. This is obviously a very primitive approximation of π. A late 18th century Rabbi, Elijah of Vilna[4] (1720–1797), was one of the great modern biblical scholars, who earned the title "Gaon of Vilna" (meaning brilliance of Vilna).

---

[1]Alfred S. Posamentier and Noam Gordon, "An Astounding Revelation on the History of π" *Mathematics Teacher*, Vo. 77, No. 1, Jan. 1984, p. 52.

[2]The "molten sea" was a gigantic bronze vessel for ritual ablutions in the court of the First Temple (966–955 B.C.). It was supported on the backs of 12 bronze oxen (volume $\approx$ 45,000 liters).

[3]A cubit is the length of a person's fingertip to his elbow.

[4]In those days Vilna was in Poland, while today the town is named Vilnius and is in Lithuania.

He came up with a remarkable discovery, one that could make most history of mathematics books faulty if they say that the Bible approximated the value of π as 3. Elijah of Vilna noticed that the Hebrew word for "line measure" was written differently in each of the two Biblical passages mentioned above.

In 1 *Kings* 7:23 it was written as הוק, whereas in 2 *Chronicles* 4:2 it was written as וק. Elijah applied the ancient biblical analysis technique (still used by talmudic scholars today) called gematria, where the Hebrew letters are given their appropriate numerical values according to their sequence in the Hebrew alphabet, to the two spellings of the word for "line measure" and found the following. The letter values are: 6 = ו, 100 = ק, and 5 = ה. Therefore, the spelling for "line measure" in 1 *Kings* 7:23 is 111 = 100 + 6 + 5 = הוק, while in 2 *Chronicles* 4:2 the spelling 106 = 100 + 6 = וק. Using gematria in an accepted way, he then took the ratio of these two values: $\frac{111}{106} = 1.0472$ (to four decimal places), which he considered the necessary "correction factor." By multiplying the Bible's apparent value (3) of π by this "correction factor," one gets 3.1416, which is π correct to four decimal places! "Wow!" is a common reaction. Such accuracy is quite astonishing for ancient times. Moreover, the teachers will remember how just getting $\pi = 3.14$ using string measurements of various circles was quite a feat. Now imagine getting π accurate to four decimal places. We would contend that this would be nearly impossible with typical string measurements. This might be a project for teachers to undertake with their students to see how close they can get to the actual value of π.

7.4. Projects

In some schools, teachers are urged to have students work on special projects involving the subject matter at hand. If this is the case, then as a coach for mathematics teachers, you might encourage and facilitate teachers to develop some projects for their students to do either in small groups or individually. A project may require collecting data from more than one class, or even throughout the school. It may even require some data obtained from home, once again involving the parents. (Note: Some of the projects may require some basic statistical manipulation and could lead to a lesson or two on the mean, median, mode, sample space, and simple correlation.) Here, we offer some project ideas that teachers might find interesting to use with their children at the appropriate grade levels. Be sure to discuss with the teachers what skills will be needed for each project, and how a particular project might be modified to make it adaptable to a variety of specific grade levels. Let's consider a few examples for various grade levels.

What is the favorite color crayon of the first-grade students in your school?

This will probably involve surveying children in several classes at the first-grade level. As a coach, you need to be well informed of any regulations that the school might have with regard to polling students throughout the school. Thus, teachers should be certain that they have permission to involve other classes.

In what month do most of your classmates have a birthday?

This project can yield an introduction to the use of bar graphs.

1. Have the teacher place the name of each month on a 5″ × 9″ card.
2. Select 12 children to each hold a month-sign.
3. Have these 12 children stand in a line in the gymnasium or other spacious place with the months in the correct order.
4. Then ask the rest of the children in the class to stand behind the month in which their birthday occurs.
5. Translate the data into a bar graph and discuss the results.

What are some of the things that your family should recycle, but actually discard?

Again, the project calls for discussion. This time it involves identifying what is recyclable and what should be discarded. It might be worthwhile to have the teachers prepare a list of some items in both categories for use by the children. This project opens the door to a discussion on recycling, care for our environment, and planet earth.

What fraction (or per cent) of the children in your grade are left-handed?

Discuss with the teachers how they might use this project with their class. Is there a large enough sample space in one class or should the project be expanded to include additional classes as well? Were the results of the sample space of the single class equal to the results of the expanded group?

How much of a banana can really be eaten?

This project would require the use of a fairly sensitive scale which might be borrowed from a science teacher. Weigh the banana as it comes from the supermarket and then weigh it again, but with the peel removed. Perhaps the fractional part should be discussed as well as the actual cost of what gets eaten. A question that will attract the children's interest is "If bananas sell for 65 cents a pound, how many pounds of super market bananas would you have to buy to obtain one pound of bananas to eat?"

Which letters in the alphabet occur most often?

For this project, give each group of children a newspaper article. Have them count each of the letters in any 10 consecutive lines of text, tallying the letters as they occur. Then arrange them in order of frequency. Here are the first thirteen actual letters of usage in order:

E T A O N I S R H L D C

How close did their sample actually come to the result shown above? Again, is there a contrast between theoretical results as opposed to the actual results? Would these results always be consistent? Discuss the concept of random sampling with the teachers.

Before you discuss introducing a specific project with their classes, have the teachers discuss the goals of the project, the mathematics it will require, and the procedures the teachers and the children might have to follow. Would the results of the project lead to some good mathematics? In the ideal situation, carefully selected projects could very nicely lead into the curriculum. This active engagement by students can also be quite motivating for the ensuing topic.

7.5. Bulletin Boards

Some schools assign bulletin boards in hallways or classrooms to various subject areas. This typically also includes mathematics. As the math coach, creating a bulletin board exhibit that attracts attention ought to be part of your responsibility — or at least assisting teachers to produce such bulletin boards which could be relevant to their instructional program. Be on the lookout for appropriate articles from newspapers and magazines. You might develop a file with folders for every grade level and at the appropriate time share these with the teachers so that they can set up a bulletin board and also take "ownership" of the board. They may well have students take an active role in developing the exhibit.

Another likely occurrence is the use of a bulletin board within the individual teacher's classroom. Most classrooms used to have a great deal of space devoted to chalkboards. With the advent of computers, calculators, white boards, smartboards, and other technology, more and more space is being given over to classroom bulletin boards. You should encourage teachers to make use of this valuable resource.

Some bulletin boards might have a place that is devoted to traditional, routine exhibits such as homework assignments for the week, notices that affect the students, and so on. However, encourage teachers to use bulletin board space to enliven the classroom, to make the room more colorful and interesting. Make mathematics realistic and attention-grabbing. For instance, one section might be devoted to a "Problem of the Week", as mentioned earlier in Chapter 1. Have the teacher put a new problem on the bulletin board on Monday morning. Students should submit their solutions as they solve the problem. When the week is over, post some of the more interesting solutions, along with the names of successful solvers.

Topics from the history of mathematics can provide many additional interesting bulletin board exhibits. For example, students might find intriguing how the early Egyptian, Roman, and Babylonian cultures represented number systems. This can be used to introduce number systems of other bases than the one we currently use. There are lots of resources not the least of which are books on the history of mathematics. As a coach, you might work with teachers either individually or in groups to get them to see the benefit of such exhibits.

7.6. Humor

There are times when comic strips in daily newspapers provide some interesting mathematics. You might want to encourage teachers to be on the look-out for such and then display them for the class to appreciate — especially when they can be made relevant to the topics being presented in class. For example, some comic strips that are widely circulated and often show cartoons with a mathematical theme are "Calvin and Hobbes" and "Foxtrot." Consider also the comic cartoons to be found at the site: https://xkcd.com, but have teachers use it selectively with the interests, level, and culture of the student population. Whenever you see any mathematics comic strip or cartoon in a newspaper or magazine, cut it out and again save it in a folder. When you have collected a number of these cartoons, you might want to put a "Cartoon of the Week" or just a simple "Math Humor" place on a hallway bulletin board. Be sure to discuss what makes the cartoon humorous and what mathematics lies behind the comic strip. Children react very positively to these since they make math fun.

The teachers you coach should design a variety of bulletin board exhibits. Have them exchange ideas among themselves. Most of all, the bulletin board should liven up the classroom.

7.7. And Finally ...

Remember, as a math coach you have many roles to play. As we have said throughout the book, you are often the liaison with the administration, keeping them advised as to what is taking place in the math classroom. The administration relies on the coach to keep them up-to-date. Needless to say, you are also directly responsible to the teachers you work with, to help them improve their teaching. You will know how well you've done when you see positive changes in their instructional program, exhibiting a wide variety of teaching strategies and a willingness by the teachers to try some alternative procedures. Remember, "change" is sometimes seen by teachers as a disturbance to their "comfort zone." Your challenge is to show that this need not necessarily be the case, but that change can be enlightening and invigorating, while continuing to make teaching fun and avoid any form of possible burnout! Indirectly, you are also responsible to the students for their mathematics success. But above all, your role is to improve the mathematics teaching and learning in your school and enhance our profession. Go for it!

www.ingramcontent.com/pod-product-compliance
Ingram Content Group UK Ltd.
Pitfield, Milton Keynes, MK11 3LW, UK
UKHW022107170926
8624IPUK00024B/194

9 789813 271708